T0314294

# COMBINED COOLING, HEATING, AND POWER SYSTEMS

# WILEY-ASME PRESS SERIES LIST

# COMBINED COOLING, HEATING, AND POWER SYSTEMS

## MODELING, OPTIMIZATION, AND OPERATION

**Yang Shi**

*University of Victoria, Canada*

**Mingxi Liu**

*University of Victoria, Canada*

**Fang Fang**

*North China Electric Power University, China*

This Work is a co-publication between ASME Press and John Wiley & Sons Ltd.

*Registered Office(s)*
John Wiley & Sons, Inc., 111 River Street, Hoboken, NJ 07030, USA
John Wiley & Sons Ltd, The Atrium, Southern Gate, Chichester, West Sussex, PO19 8SQ, UK

*Editorial Office*
The Atrium, Southern Gate, Chichester, West Sussex, PO19 8SQ, UK

For details of our global editorial offices, customer services, and more information about Wiley products visit us at www.wiley.com.

Wiley also publishes its books in a variety of electronic formats and by print-on-demand. Some content that appears in standard print versions of this book may not be available in other formats.

*Library of Congress Cataloging-in-Publication Data*

Names: Shi, Yang, 1972- author. | Liu, Mingxi, 1988- author. | Fang, Fang, 1976- author.
Title: Combined cooling, heating, and power systems : modeling, optimization, and operation / Yang Shi, Mingxi Liu, Fang Fang.
Description: Singapore ; Hoboken, NJ : John Wiley & Sons, 2017. | Includes bibliographical references and index. | Description based on print version record and CIP data provided by publisher; resource not viewed.
Identifiers: LCCN 2017004053 (print) | LCCN 2017012538 (ebook) | ISBN 9781119283379 (Adobe PDF) | ISBN 9781119283423 (ePub) | ISBN 9781119283355 (cloth)
Subjects: LCSH: Cogeneration of electric power and heat. | Cooling systems. | Heating.
Classification: LCC TK1041 (ebook) | LCC TK1041 .S55 2017 (print) | DDC 621.1/99–dc23
LC record available at https://lccn.loc.gov/2017004053

Cover image: © artJazz/Gettyimages
Cover design by Wiley

Set in 10/12pt, TimesLTStd by SPi Global, Chennai, India
Printed and bound in Malaysia by Vivar Printing Sdn Bhd

10 9 8 7 6 5 4 3 2 1

*To my beloved parents and family*
*–Yang Shi*

*To my beloved parents and Jingwen*
*–Mingxi Liu*

*To my beloved parents and family*
*–Fang Fang*

# Contents in Brief

# Contents

# List of Figures

# List of Tables

# Series Preface

The Wiley-ASME Press Series in Mechanical Engineering brings together two established leaders in mechanical engineering publishing to deliver high-quality, peer-reviewed books covering topics of current interest to engineers and researchers worldwide. The series publishes across the breadth of mechanical engineering, comprising research, design and development, and manufacturing. It includes monographs, references and course texts.

Prospective topics include emerging and advanced technologies in Engineering Design; Computer-Aided Design; Energy Conversion & Resources; Heat Transfer; Manufacturing & Processing; Systems & Devices; Renewable Energy; Robotics; and Biotechnology.

# Preface

Combined cooling, heating and power (CCHP) is a feature of trigeneration systems able to supply cooling, heating, and electricity simultaneously. CCHP systems can be employed to provide buildings with cooling, heating, electricity, hot water and other uses of thermal energy. CCHP features with the great potential of dramatically increasing resource energy efficiency and reducing carbon dioxide emissions. Our intention through this book is to provide a timely account as well as an introductory exposure to the main developments in modeling, optimization, and operation of CCHP systems. At the time of conceiving this project, we believed that the development of a systematic framework on modeling and optimal operation design of CCHP systems was of paramount importance. A concise overview of the research area is presented in Chapter 1. We hope it will help readers arrive at a broader and more balanced view of CCHP systems. The remainder of the book presents the core contents, which are divided into five chapters. In Chapter 2, based on two conventional operation strategies, that is, following electric load (FEL) and following thermal load (FTL), a novel optimal switching operation strategy is presented. Chapter 3 presents a configuration with hybrid chillers and design of the optimal operation strategy. In Chapter 4, based on the concept of energy hub, a system matrix-based model is proposed to systematically facilitate the design of optimal operation strategies. Chapter 5 discusses the load prediction problem which plays an instrumental role in designing CCHP operation schemes. In Chapter 6, a complementary CCHP-organic Rankine cycle (CCHP-ORC) system is introduced.

The writing of this monograph has benefitted greatly from discussions with many colleagues. We wish to express our heartfelt gratitude to Professor Jizhen Liu who shared many of his ideas and visions with us. Others who contributed directly by means of joint research on the subject include Le Wei, Qinghua Wang, Hui Zhang, and Huiping Li, with whom we have enjoyed many collaborations. We have also benefitted from constructive and enlightening discussions with Jianhua Zhang, Guolian Hou, Jian Wu, Ji Huang, Xiaotao Liu, Chao Shen, Yuanye Chen, Bingxian Mu, Jicheng Chen, and Kunwu Zhang, among others. Support from the Natural Sciences and Engineering Research Council of Canada, from the National Natural Science Foundation of China (under grant 61473116 and 51676068) has been very helpful and is gratefully acknowledged. Finally, as a way of expressing our deep gratitude and indebtedness, the first author dedicates this book to his wife Jing, and Eric and Adam, the second author to his wife Jingwen, and the third author to his wife Le, and Bowen and Yihe, for their great support and encouragement on this project.

Yang Shi, Mingxi Liu, Fang Fang
*Victoria, BC, Canada*

# Acknowledgment

The authors would like to thank all those who have helped in accomplishing this book.

# Acronyms

| | |
|---|---|
| AFC | Alkaline Fuel Cell |
| ANN | Artificial Neural Network |
| AR | AutoRegressive |
| ARIMA | AutoRegressive Integrated Moving Average |
| ARMA | AutoRegressive Moving Average |
| ARMAX | AutoRegressive Moving Average with eXogenous inputs |
| ATC | Annual Total Cost |
| ATCS | Annual Total Cost Saving |
| ATD | Aggregate Thermal Demand |
| BFGS | Broyden–Fletcher–Goldfarb–Shanno |
| CCHP | Combined Cooling, Heating, and Power |
| CDE | Carbon Dioxide Emissions |
| CDER | Carbon Dioxide Emissions Reductions |
| CHP | Combined Heating and Power |
| CITHR | Cooling-side Incremental Trigeneration Heat Rate |
| COP | Coefficient of Performance |
| DHC | District Heating and Cooling |
| DOE | Department of Energy |
| EA | Evolutionary-Algorithmic |
| EBMUD | East Bay Municipal Utility District |
| EC | Evaluation Criteria |
| EDM | Electric Demand Management |
| EITHR | Electrical-side Incremental Trigeneration Heat Rate |
| EPA | Environmental Protection Agency |
| EUETS | European Union Emissions Trading Scheme |
| ec | Electric Chiller |
| FCL | Following Constant Load |
| FEL | Following the Electric Load |
| FTL | Following the Thermal Load |
| GA | Genetic Algorithm |
| GHG | GreenHouse Gas |
| GRG | Generalized Reduced Gradient |
| GRU | Gainsville Regional Utilities |

| | |
|---|---|
| HETL | Hybrid Electric-Thermal Load |
| hrc | Recovered Heat for Cooling |
| hrh | Recovered Heat for Heating |
| HRSG | Heat Recovery Steam Generator |
| hrs | Heat Recovery System |
| HTC | Hourly Total Cost |
| HTCS | Hourly Total Cost Savings |
| HVAC | Heating, Ventilation, and Air Conditioning |
| IC | Internal Combustion |
| IV | Instrument Variable |
| KKT | Karush–Kuhn–Tucker |
| LP | Linear Programming |
| LS | Least Squares |
| MA | Moving Average |
| MAE | Mean Absolute Error |
| MAFC | Magnesium-Air Fuel Cell |
| MAPE | Mean Absolute Percentage Error |
| MCFC | Molten Carbonate Fuel Cell |
| MILP | Mixed Integer Linear Programming |
| MINLP | Mixed Integer Non-Linear Programming |
| MSPE | Mean Square Prediction Error |
| MPC | Model Predictive Control |
| OLS | Ordinary Least Squares |
| ORC | Organic Rankine Cycle |
| PAFC | Phosphoric Acid Fuel Cell |
| PEMFC | Proton Exchange Membrane Fuel Cell |
| PEC | Primary Energy Consumption |
| PES | Primary Energy Savings |
| PGU | Power Generation Unit |
| PURPA | Public Utility Regulatory Policy Act |
| PV | PhotoVoltaic |
| QP | Quadratic Programming |
| SNPV | System Net Present Value |
| SOFC | Solid Oxide Fuel Cell |
| SP | Separation Production |
| SQP | Sequential Quadratic Programming |
| TDM | Thermal Demand Management |
| TITHR | Thermal-side Incremental Trigeneration Heat Rate |
| TPES | Trigeneration Primary Energy Saving |
| TRR | Total Revenue Requirement |
| TSLS | Two-Stage Least Squares |
| TSRLS | Two-Stage Recursive Least Squares |
| WADE | World Alliance for Decentralized Energy |

# Symbols

| | |
|---|---|
| $a_i(\star)$ | The $i$th equality constraint of variable $\star$ |
| $ATC$ | Annual total cost |
| $ATCS$ | Annual total cost savings |
| $C_{ac}$ | Unit price of the absorption chiller |
| $C_b$ | Unit price of the boiler |
| $C_{ca}$ | Carbon tax rate |
| $C_e$ | Electricity rate |
| $C_{ec}$ | Unit price of the electric chiller |
| $C_f$ | Natural gas rate |
| $C_h$ | Unit price of the heating unit |
| $c_j(\star)$ | The $j$th inequality constraint of variable $\star$ |
| $C_{pgu}$ | Unit price of the PGU |
| $C_s$ | Electricity sold-back rates |
| $CDE$ | Carbon dioxide emissions |
| $CDE^{CCHP}$ | Carbon dioxide emissions of the CCHP system |
| $CDE^{CCHP}_{FEL}$ | Carbon dioxide emissions of the CCHP system under FEL |
| $CDE^{CCHP}_{FTL}$ | Carbon dioxide emissions of the CCHP system under FTL |
| $CDE^{SP}$ | Carbon dioxide emissions of the SP system |
| $CDER$ | Carbon dioxide emissions reductions |
| $COP_{ac}$ | Coefficient of performance of the absorption chiller |
| $COP_{ec}$ | Coefficient of performance of the electric chiller |
| $COST$ | Operational cost |
| $COST^{CCHP}_{FEL}$ | Operational cost of the CCHP system under FEL |
| $COST^{CCHP}_{FTL}$ | Operational cost of the CCHP system under FTL |
| $COST^{SP}$ | Operational cost of the SP system |
| $Cov[\bullet, \star]$ | Covariance of variables $\bullet$ and $\star$ |

| | |
|---|---|
| $E[\star]$ | Expectation of variable $\star$ |
| $E_{ec}$ | Electricity consumed by the electric chiller in the CCHP system |
| $E_{ec}^{SP}$ | Electricity consumed by the electric chiller in the SP system |
| $E_{excess}$ | Excess electricity |
| $E_{grid}$ | Purchased electricity from the grid by the CCHP system |
| $\check{E}_{grid}(t)$ | Purchased electricity for compensating for the cooling gap |
| $E_{grid}^{SP}$ | Purchased electricity from the grid by the SP system |
| $e_i$ | Standard basis vector with the $i$th element being 1 |
| $E_i^{\ell}$ | Electricity input of component $\ell$ |
| $E_o^{\ell}$ | Electricity output of component $\ell$ |
| $\bar{E}_o^{pgu}$ | Maximum electricity generated by the PGU |
| $E_{orc}$ | Electricity output of the ORC |
| $E_p$ | Parasitic electricity |
| $E_{pgu}$ | Electricity generated from the PGU |
| $\bar{E}_{pgu}$ | Maximum electricity generated by the PGU |
| $E_{pgu-FEL}$ | Electricity generated from the PGU under FEL |
| $E_{pgu-FTL}$ | Electricity generated from the PGU under FTL |
| $E_{pro}$ | Electricity generated by the PGU |
| $E_{req}$ | Electricity required by building users and the electric chiller |
| $E_{user}$ | Electricity required by building users |
| $E_{userl}$ | Lower bound of electricity required by building users |
| $E_{useru}$ | Upper bound of electricity required by building users |
| $EC$ | Evaluation criteria function value |
| $EC_{annual}$ | Annual evaluation criteria function value |
| $EC_{FEL}$ | Evaluation criteria function value of the CCHP system under FEL |
| $EC_{FTL}$ | Evaluation criteria function value of the CCHP system under FTL |
| $EC_{hour}$ | Hourly evaluation criteria function value |
| $EC_{hour,ij}$ | Hourly evaluation criteria function value of day $i$, hour $j$ |
| $F_b$ | Fuel consumed by the boiler in the CCHP system |
| $F_b^{SP}$ | Fuel consumed by the boiler in the SP system |
| $F_{b-FEL}$ | Fuel consumed by the boiler in the CCHP system under FEL |
| $F_{b-FTL}$ | Fuel consumed by the boiler in the CCHP system under FTL |
| $F^{CCHP}$ | Fuel consumed by the CCHP system |
| $F_i^{\ell}$ | Fuel input of component $\ell$ |
| $F_m$ | Total fuel consumption |
| $\check{F}_m$ | Additionally purchased fuel |

| | |
|---|---|
| $F_{m-FEL}$ | Total fuel consumption of the CCHP system under FEL |
| $F_{m-FTL}$ | Total fuel consumption of the CCHP system under FTL |
| $F_o^\ell$ | Fuel output of component $\ell$ |
| $F_{pgu}$ | Fuel consumed by the PGU |
| $F_{pgu-FEL}$ | Fuel consumed by the PGU in the CCHP system under FEL |
| $F_{pgu-FTL}$ | Fuel consumed by the PGU in the CCHP system under FTL |
| $F_{pgum}$ | Maximum fuel consumption of the PGU |
| $F_{pgumopt}$ | Optimal PGU capacity |
| $F_{red}$ | Reduced fuel consumption |
| $F^{SP}$ | Fuel consumed by the SP system |
| $\mathcal{H}^\ell$ | Energy conversion matrix of component $\ell$ |
| $h_1$ | Enthalpy of organic fluid at the inlet of pump |
| $h_2$ | Enthalpy of organic fluid at the outlet of pump |
| $h_{2s}$ | Enthalpy at the outlet of pump for the isentropic case |
| $h_3$ | Enthalpy of organic fluid at the outlet of the evaporator |
| $h_4$ | Enthalpy of organic fluid at the outlet of the pump |
| $h_{4s}$ | Enthalpy of organic fluid at the outlet of the turbine for the isentropic case |
| $HTC$ | Hourly total cost |
| $HTC^{CCHP}$ | Hourly total cost of the CCHP system |
| $HTC^{SP}$ | Hourly total cost of the SP system |
| $HTCS$ | Hourly total cost savings |
| $K$ | Power to heat ratio |
| $k_e$ | Site-to-primary energy conversion factor for electricity |
| $k_f$ | Site-to-primary energy conversion factor for natural gas |
| $L$ | Facility's life |
| $\max f(\bullet)$ | Maximize the function value of $f(\bullet)$ |
| $\min f(\bullet)$ | Minimize the function value of $f(\bullet)$ |
| $\max\{\bullet, \star\}$ | Maximum value between $\bullet$ and $\star$ |
| $\min\{\bullet, \star\}$ | Minimum value between $\bullet$ and $\star$ |
| $m_{orc}$ | Organic fluid mass flow rate |
| $PEC$ | Primary energy consumption |
| $PEC^{CCHP}$ | Primary energy consumption of the CCHP system |
| $PEC_{FEL}^{CCHP}$ | Primary energy consumption of the CCHP system under FEL |
| $PEC_{FTL}^{CCHP}$ | Primary energy consumption of the CCHP system under FTL |
| $PEC^{SP}$ | Primary energy consumption of the SP system |

| $PES$ | Primary energy savings |
|---|---|
| $Q_{ac}$ | Cooling energy provided by the absorption chiller |
| $Q_c$ | Total cooling demand |
| $Q_{cd}$ | Heat exchange of the condenser |
| $Q_{ec}$ | Cooling energy provided by the electric chiller |
| $Q_{ep}$ | Obtained heat by evaporator |
| $Q_{eq}$ | Equivalent total thermal requirement at the output of the heat recovery system |
| $Q_b$ | Thermal energy provided by the boiler in the CCHP system |
| $Q_b^{SP}$ | Thermal energy provided by the boiler in the SP system |
| $Q_{gap}$ | Thermal energy gap |
| $Q_h$ | Total heating demand |
| $Q_{hi}^{\ell}$ | Heating input of component $\ell$ |
| $Q_{ho}^{\ell}$ | Heating output of component $\ell$ |
| $Q_{hrc}$ | Thermal energy from the heat recovery system for the use of cooling |
| $Q_{hrh}$ | Thermal energy from the heat recovery system for the use of heating |
| $Q_{pro}$ | Thermal energy provided by the PGU |
| $Q_r$ | Thermal energy provided by the heat recovery system |
| $Q_{req}$ | Thermal energy required by building users and the electric chiller |
| $Q_{r-FEL}$ | Thermal energy provided by the heat recovery system under FEL |
| $Q_{r-FTL}$ | Thermal energy provided by the heat recovery system under FTL |
| $Q_{ro}$ | Thermal input of the ORC |
| $Q_{user}$ | Total thermal demand by building users |
| $R$ | Capital recovery factor |
| $T_{dew}$ | Dew-point temperature |
| $T_{dew}^{o}$ | Observation of the dew-point temperature |
| $T_{dry}$ | Dry-bulb temperature |
| $T_{dry}^{o}$ | Observation of the dry-bulb temperature |
| $\hat{T}_{dry}$ | Estimation of the dry-bulb temperature |
| $\mathcal{V}_i^{\ell}$ | Energy input vector of component $\ell$ |
| $\mathcal{V}_o^{\ell}$ | Energy output vector of component $\ell$ |
| $\hat{\mathcal{V}}_o$ | Forecasted load vector |
| $\bar{\mathcal{V}}_o^{\ell}$ | Upper bound of the output of component $\ell$ |
| $\underline{\mathcal{V}}_o^{\ell}$ | Lower bound of the output of component $\ell$ |
| $\mathrm{Var}[\star]$ | Variance of variable $\star$ |
| $W_p$ | Pump power |

| | |
|---|---|
| x | Electric cooling to cool load ratio |
| $y_c$ | Variable of cooling load |
| $\hat{y}_c$ | Variable of forecasted cooling load |
| $\tilde{y}_c$ | Variable of remained cooling to be provided |
| $y_e$ | Variable of electric load |
| $\hat{y}_e$ | Variable of forecasted electric load |
| $y_h$ | Variable of heating load |
| $\hat{y}_h$ | Variable of forecasted heating load |
| $\tilde{y}_h$ | Variable of remained heating to be provided |
| $z^{-\star}$ | $\star$ time lags from the current time instant |
| $\Gamma_\ell$ | Dispatch matrix of component $\ell$ |
| $\eta_h$ | Efficiency of the heating unit |
| $\eta_{pgu}$ | Efficiency of the PGU |
| $\eta_{hrs}$ | Efficiency of the heat recovery system |
| $\eta_b$ | Efficiency of the boiler |
| $\eta_e^{SP}$ | Generation efficiency of the SP system |
| $\eta_{grid}$ | Transmission efficiency of local grid |
| $\eta_p$ | Isentropic efficiency |
| $\eta_{orc}$ | Efficiency of the ORC |
| $\eta_{gen}$ | Efficiency of the electric generator |
| $\mu_e$ | Carbon dioxide emissions conversion factor of electricity |
| $\mu_f$ | Carbon dioxide emissions conversion factor of natural gas |
| $\xi$ | Evaporator effectiveness |
| $\omega_i$ | Weighting coefficient of the $i$th criterion |
| $\nabla$ | Gradient |
| $^\circ C$ | Centigrade |
| $\exists$ | Exists |
| $\in$ | In |
| $\triangleq$ | Define |
| $\sum$ | Sum |
| $\forall$ | For all |
| s.t. | Subject to |
| $\top$ | Matrix/vector transpose |
| $\mathbb{R}^n$ | Real vector space of dimension $n$ |
| $\mathbb{R}^{n \times m}$ | Real matrix space of dimension $n \times m$ |
| $\bullet^*$ | The optimal value of variable $\bullet$ |
| $O$ | Complexity |

# Introduction

Combined cooling, heating, and power (CCHP) systems are known as trigeneration systems. They are designed to supply cooling, heating, and electricity simultaneously. The CCHP system has become a hot topic for its high system efficiency, high economic efficiency, and low greenhouse gas (GHG) emissions in recent years. The efficiency of the CCHP system depends on the appropriate system configuration, operation strategy, and facility selection. Due to the inherent and inevitable energy waste of traditional operation strategies, high-efficiency operation strategies are urged. To achieve the highest system efficiency, facilities in the system should be appropriately sized to match with the corresponding operation strategy.

In Chapter 1, the state-of-the-art of CCHP research is surveyed. First, the development and working scheme of the CCHP system is presented. Some analyses of the advantages of this system and a brief introduction to the related components are then given. In the second part of Chapter 1, we elaborately introduce various types of prime movers and thermally activated facilities. Recent research progress on the management, control, system optimization, and facility selection is summarized in the third part. The development of the CCHP system in representative countries and the development barriers are also discussed in Chapter 1.

The operation strategy has a direct impact on the CCHP system performance. To improve the operational performance, in Chapter 2, based on two conventional operation strategies, that is, following electric load (FEL) and following thermal load (FTL), a novel optimal switching operation strategy is proposed. Using this strategy, the whole operating space of the CCHP system is divided into several regions by one to three border surfaces determined by energy requirements and the evaluation criteria (EC). Then the operating point of the CCHP system is located in a corresponding operating mode region to achieve improved EC. The EC simultaneously considers the primary energy consumption, the operational cost, and the carbon dioxide emissions. The proposed strategy can reflect and balance the influences of energy requirements, energy prices, and emissions effectively.

Most of the improved operation strategies in the literature are based on the "balance" plane, matching of the electric demands with the thermal demands. However, in more than 95% energy demand patterns, the demands cannot match with each other on this exact "balance" plane. To continuously use the "balance" concept, in Chapter 3, the system configuration is modified from the one with a single absorption chiller to be the one with hybrid chillers, thus expanding the "balance" plane to a "balance" space by tuning the electric cooling to cool load ratio. With this new "balance" space, an operation strategy is designed and the power generation unit (PGU) capacity is optimized according to the proposed operation strategy to reduce the energy waste and improve the system efficiency. A case study is conducted to verify the feasibility and effectiveness of the proposed operation strategy.

In Chapter 4, a more mathematical approach to scheduling the energy input and power flow is proposed. By using the concept of *energy hub*, the CCHP system is modeled in a matrix form. As a result, the whole CCHP system is an input–output model. Setting the objective function to be a weighted summation of primary energy savings (PES), hourly total cost savings (HTC), and carbon dioxide emissions reductions (CDER), the optimization problem, constrained by equality and inequality constraints, is solved to obtain the optimal operation strategy. The PGU capacity is also sized under the proposed optimal operation strategy. In the case study, compared with FEL and FTL, the proposed optimal operation strategy saves more primary energy and annual total cost, and can be more environmentally friendly.

Most of the current operation strategies are designed by assuming that accurate loads during the next time interval are already known. In Chapter 5, in order to solve the problem of unknown loads in practical applications, by using an AutoRegressive Moving Average with eXogenous inputs (ARMAX) model, whose parameters are identified by a proposed Ordinary Least Squares–Two-Stage Recursive Least Squares (OLS-TSRLS) algorithm, cooling, heating, and electrical loads in the future time intervals are forecasted. The identification procedure uses the dew-point temperature as the instrumental variable (IV) for the exogenous variable (dry-bulb temperature) to better explain the relation between exogenous and endogenous variables. TSRLS at the second stage helps to reduce the time complexity. A post-strategy is also proposed to compensate for the inaccurate forecasting. A case study is conducted to verify the feasibility and effectiveness of the proposed methods.

The electricity to thermal energy output ratio is an important impact factor for the operation strategy and performance of CCHP systems. If the energy requirements of users are managed to just match this ratio, the system efficiency would reach the maximum. However, due to the randomness of users' demand, this situation is rarely achieved in practice. To solve this problem, a complementary CCHP-organic Rankine cycle (CCHP-ORC) system is configured in Chapter 6. The salient feature of this system is that its electricity to thermal energy output ratio can be adjusted by changing the loads of the electric chiller and the ORC dynamically. For such a system, an optimal operation strategy and a corresponding implemented decision-making process are presented within a wide load range. Case studies are conducted to verify the efficacy of the developed CCHP-ORC system.

# 1

# State-of-the-Art of Combined Cooling, Heating, and Power (CCHP) Systems

## 1.1 Introduction

With the rapid development of distributed energy supply systems [1, 2, 3, 4], combined heating and power (CHP) systems and combined cooling, heating, and power (CCHP) systems have become the core solutions to improve the energy efficiency and to reduce greenhouse gas (GHG) emissions [5, 6, 7, 8, 9]. The CCHP system is an extended concept of the CHP system, which has been widely utilized in large-scale centralized power plants and industrial applications [10]. CHP systems are developed to conquer the problem of low energy efficiency of conventional separation production (SP) systems. In SP systems, electric demands, which include daily electricity usage and electric chiller usage, and heating demands are provided by the purchased electricity and fuel, respectively. Since no self-generation exists in SP systems, they are proved to be of low efficiency; however, in CHP systems, most of the electric and heating demands are provided simultaneously by a prime mover together with a heat recovery system, a heat storage system, and so on. Energy demands beyond the system capacity can be supplied by the local grid and an auxiliary boiler. If some thermally activated technologies are introduced, for example, absorption and adsorption chillers, into the CHP to provide the cooling energy, the original CHP system evolves to a CCHP system [11], which can also be referred to as a *trigeneration* system and building cooling heating and power (BCHP) system. Since there is no cooling need in winter, the CHP system can be regarded as a special case of the CCHP system. A CCHP system can achieve up to 50% greater system efficiency than a CHP plant of the same size [12].

*Combined Cooling, Heating, and Power Systems: Modeling, Optimization, and Operation,* First Edition.
Yang Shi, Mingxi Liu, and Fang Fang.
© 2017 John Wiley & Sons Ltd. Published 2017 by John Wiley & Sons Ltd.

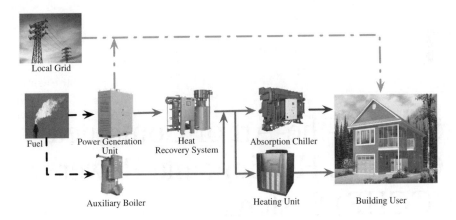

**Figure 1.1**   A typical CCHP system

A typical CCHP system is shown in Figure 1.1. The power generation unit (PGU) provides electricity for the user. Heat, produced as a by-product, is collected to meet cooling and heating demands via the absorption chiller and heating unit. If the PGU cannot provide enough electricity or by-product heat, additional electricity and fuel need to be purchased to compensate for the electric gap and feed the auxiliary boiler, respectively. In this way, three types of energy, that is, cooling, heating, and electricity, can be supplied simultaneously.

Compared with conventional generating plants, the advantages of a CCHP system are three-fold: high efficiency, low GHG emissions, and high reliability.

First, the high overall efficiency of a CCHP system implies that less primary fuel is consumed in this system to obtain the same amount of electric and thermal energy. In [10], the authors give an example to show that, compared with the traditional energy supply mode, the CCHP system can improve the overall efficiency from 59% to 88%. This improvement owes to the cascade utilization of different energy carriers and the adoption of the thermally activated technologies. As the main electricity source, the PGU has an electric efficiency as low as 30%. By implementing the heat recovery system, the CCHP system can collect the by-product heat to feed the absorption/adsorption chiller and heating unit to provide cooling and heating energy, respectively. By adopting the absorption chiller, no additional electricity needs to be purchased from the local grid to drive the electric chiller in summer, but only the recovered heat is used. In winter, a CCHP system degenerates to be a CHP system. The high efficiency of the CHP system is investigated in [13, 14, 15, 16, 17, 18, 19, 20]. In a nutshell, a CCHP system can dramatically reduce the primary consumption and improve the energy efficiency.

The second advantage involved in the CCHP system is the low GHG emissions. On the one hand, the trigeneration structure of the CCHP system contributes to this reduction. Compared with SP systems, if within the capacity limitation of the prime mover, no additional electricity needs to be purchased from the local grid, which is supplied by fossil-fired power plants. It is well known that, even though the penetration of some types of renewable energy, for example, the wind, tide and solar energy, increase significantly [21, 22, 23], because of their intermittency, the main electricity producer is still the fossil-fired power plant. By reducing the consumption of electricity from the local grid, GHG emissions from fossil-fired power plants can be decreased. Moreover, adopting the thermally activated technologies can also reduce the electricity consumption by the electric chiller, which will result in less consumption of fossil fuel in the grid power plant. On the other hand, new technologies in the prime mover also contribute to the GHG emissions reduction. Incorporating fuel cells, which are one of the hottest topics in recent years, in the CCHP system can increase the system efficiency up to 85–90% [24]. Compared with some conventional prime movers, such as the internal combustion (IC) engine and combustion turbine, the new-tech prime movers can provide the same amount of electricity with less fuel supply and less GHG emissions. In recent years, aiming to reduce GHG emissions, an increasing number of countries have begun to run the carbon tax act [25, 26, 27, 28, 29]. As a result of these acts, reducing GHG emissions can not only reduce the contaminant of the air, but also can improve the system's economic efficiency.

The other benefit brought by the CCHP system is reliability, which can be regarded as the ability to guarantee the energy supply at a reasonable price [30]. Recent cases have demonstrated that centralized power plants are vulnerable to natural disasters and unexpected phenomena [31]. Changes in climate, terrorism, customer needs, and the electricity market are all fatal threats to the centralized power plants [10]. The CCHP system, which adopts the distributed energy technologies, can be resistant to external risks and has no electricity blackouts, for it is independent of electricity distribution. A comparison of the reliability between the distributed and centralized energy systems in Finland and Sweden can be found in [30].

A typical CCHP system consists of a PGU, a heat recovery system, thermally activated chillers, and a heating unit. Normally, the PGU is a combination of a prime mover and an electricity generator. The rotary motion generated by the prime mover can be used to drive the electricity generator. There are various options for the prime mover, for example, steam turbines, stirling engines, reciprocating IC engines, combustion turbines, micro-turbines, and fuel cells. The selection of the prime mover depends on current local resources, system size, budget limitation and GHG emissions policy. The heat recovery system plays a role in collecting the by-product heat from the prime mover. The most frequently used thermally activated technology in the CHP/CCHP system is the absorption chiller. Some novel solutions, such as the adsorption chiller, and the hybrid chiller, are also adopted in CCHP systems [32, 33, 34, 35, 36]. The selection of the heating unit depends on the design of the heating, ventilation and air conditioning (HVAC) components.

With the benefits of high system and economic efficiency, and less GHG emissions, CCHP systems have been widely installed in hospitals, universities, office buildings, hotels, parks, supermarkets, and so on [37, 38, 39, 40, 41]. For example, in China, the CCHP project at Shanghai Pudong International Airport generates combined cooling, heating, and electricity for the airport's terminals at peak demand times. It is fuelled by natural gas from offshore in the East China Sea [42]. This system is equipped with one 4 MW natural gas turbine, one 11 t/h waste heat boiler, cooling units of four YORK OM 14 067 kW, two YORK 4220 kW, four 5275 kW steam LiBr/water chillers, three 30 t/h gas boilers and one 20 t/h as standby for heat supply [43]. In the last decade, the installation of CCHP systems has plateaued. Especially, the development is much slower in developing countries than that in developed countries due to the following barriers: less public awareness, insufficient incentive policies and instruments, non-uniform design standards, incomplete connections with power grid, high price and supply pressure of natural gas, and difficulties in manufacturing equipment [43]. According to a survey by the World Alliance for Decentralized Energy (WADE), the penetration of CCHP systems can be increased by introducing the European Union Emissions Trading Scheme (EUETS) and increasing carbon tax.

This chapter aims to provide some fundamental information and the state-of-the-art of CCHP systems. Analyses and comparisons of system components, suitability scope, operating economy, system configurations and operation strategies are given for the purpose of engineering assessment. This chapter is organized as follows: in Section 1.2, different prime movers for driving the CCHP systems are introduced and compared; three main thermally activated technologies that can be used in CCHP systems to achieve energy cascade utilization are introduced in Section 1.3; Section 1.4 focuses on different system configurations according to the system capacity; in Section 1.5, conventional and novel operation strategies, and system optimization methods are introduced, analyzed and compared; development of CCHP systems in three main countries are discussed in Section 1.6; and Section 1.7 concludes this chapter.

## 1.2  Prime Movers

A prime mover, defined as a machine that transforms energy from thermal, electrical or pressure form to mechanical form, typically an engine or turbine, is the heart of an energy system. Normally, the output of a prime mover is the rotary motion, so it is always being used to couple with an electric generator. In recent years, the prime movers that are installed the most are gas engines and gas turbines [44]. These two types of prime movers belong to the reciprocating IC engine and the combustion turbine/micro-turbine, respectively. Some other types of prime movers, such as steam turbines, micro-turbines, stirling engines and fuel cells, are also being used in CCHP systems in some particular cases. In this section, emphasis will be put on reciprocating IC engines and combustion turbines/micro-turbines; other types will also be discussed.

### 1.2.1  Reciprocating IC Engines

A reciprocating engine, also known as a piston engine, is a heat engine that uses one or more reciprocating pistons to convert pressure into a rotary motion [45]. There exist two types of reciprocating engines, that is, spark ignition, which uses the natural gas as the preferred fuel and can also be fed by the propane, gasoline or landfill gas, and compress ignition, which can operate on diesel fuel or heavy oil [46]. The size of the reciprocating engines can range from 10 kW to over 5 MW.

As stated in [47, 48], with the advantages of low capital cost, quick starting, good load following performance, relatively high partial load efficiency and generally high reliability, the reciprocating engines have been widely used in many distributed generation applications, such as the industrial, commercial and institutional facilities for power generation, and CCHP systems. The waste heat of the reciprocating engine, consisting of exhaust gas, engine jacket water, lube oil cooling water and turbocharger cooling [46], can be used in thermally activated facilities in the CCHP system. This energy cascade utilization can efficiently improve the system's efficiency. However, a reciprocating engine does need regular maintenance and service to ensure its reliability [49]. Since the rising level of GHG emissions has become a big concern, applications of diesel fueled engines are restricted for the high emission level of $NO_x$. Current natural gas ignition engines have relatively low emissions profiles and are widely installed. One example is where HONDA has developed a new cogenerator, which is a natural gas-powered engine, powered by "GF160V". This cogeneration unit can reduce the $CO_2$ emissions up to approximately 20% [50]. Another classical application of the reciprocating engine example is the CHP plant with IC engines which has been installed at the Faculty of Engineering of the University of Perugia since 1994 [51]. The experimental results show that by introducing the absorption cooler into the plant can dramatically reduce the payback period.

Reciprocating IC engines are quite popular in some applications when working together with an electric or absorption chiller. High temperature exhaust gas from the engine can be used to provide heating and cooling, or to drive the desiccant dehumidifier. Maidment and Tozer [40] analyze a CCHP system for a typical supermarket using a gas turbine and a LiBr/water absorption chiller. They discuss the methodology for choosing the prime mover between a gas engine and a gas turbine. The result shows that this CCHP system offers significant primary energy consumption savings and $CO_2$ savings compared with conventional heat and power schemes. In [52], the authors assess a CCHP system, which is driven by a reciprocating IC engine, combined with a desiccant cooling system. This system incorporates a desiccant dehumidifier, a heat exchanger, and a direct evaporative cooler. The parametric analysis provided in this work shows that combining the desiccant cooling system can handle both latent and sensible loads in a wide range of climate conditions. The coefficient of performance (COP) of this system is 1.5 times that of the conventional system. Longo et al. [53] discuss a CCHP system equipped with an Otto engine and an absorption machine. The exhaust thermal energy is recovered to drive a double-effect LiBr/water cycle, and the heat recovered from the cooling jacket is used to drive a single-effect LiBr/water cycle. In [54], Talbi and Agnew explore the theoretical performance of four different configurations of a CCHP system equipped with a turbocharger diesel engine and an absorption refrigeration unit. The results show the potential of using a diesel absorption combined cycle with pre-inter cooling to achieve higher power output and thermal efficiency among other configurations. The situation of $CO_2$ emissions of CCHP systems with gas engines under different working conditions is discussed in [55].

## 1.2.2   Combustion Turbines

The combustion turbine, also known as the gas turbine, is an engine in which the combustion of a fuel, usually the gas, occurs with an oxidizer in a combustion chamber [56]. Combustion turbines have been used for the purpose of generating electricity since the 1930s. The size of gas turbines ranges from 500 kW to 250 MW, which makes it suitable for large-scale cogeneration or trigeneration systems. At partial load, the efficiency of the gas turbine can be unacceptably lower than full-capacity efficiency. As a result, generation sets smaller than 1 MW are proven to be uneconomical [10]. Gas turbines also produce high-quality (high-temperature around 482°C) exhaust heat that can be used by thermally activated processes in CCHP systems to produce cooling, heating, or drying, and to raise the overall system efficiency to approximately 70–80% [57]. Adopting some cycle integration technologies, such as steam injection gas turbines and humid air turbines, can improve the performance of the simple-cycle gas turbine by integrating the bottoming water/steam cycle into the gas turbine cycle in the form of water or steam injection [58].

For GHG emissions, because of the use of natural gas, when compared with other liquid or solid fuel-fired prime movers, gas turbines can dramatically reduce $CO_2$ emissions per kilowatt-hour [59]. Emissions of $NO_x$ can be below 25 ppm and CO emissions can be in the range of 10–50 ppm. Some emission control approaches, such as the diluent injection, lean premixed combustion, selective catalytic reduction, carbon monoxide oxidation catalysts, catalytic combustion and catalytic absorption systems can also help to reduce $NO_x$ emissions.

One typical application of gas-turbine-based cogeneration or trigeneration systems is for colleges or university campuses, where the produced steam is used to provide space heating in winter and cooling in summer. Another typical application is for the supermarket. In the USA, CCHP systems have been widely installed in supermarkets to improve the system efficiency. Produced steam and heat from the gas turbine is used to drive the food-refrigeration system, which requires a huge amount of cooling energy, and to provide the basic space heating [60]. CCHP systems using gas turbines have attracted a certain amount of attention. Exergy analyses for a combustion-gas-turbine-based power generation system are addressed in [61], which can be used for engineering design and component selection. Investigations of CCHP systems using gas turbines can be found in [7, 62]. In the latter, a micro-CCHP system with a small gas engine and absorption chiller is built in Shanghai Jiao Tong University with the designed energy management method, which can also be used in large-scale CCHP systems whose overall efficiency can be as high as 76%.

## 1.2.3   Steam Turbines

A steam turbine is a mechanical device that extracts thermal energy from pressurized steam, and converts it into rotary motion [63]. Compared with reciprocating steam engines, the higher efficiency and lower cost make steam turbines have been used for about 100 years. The size of steam turbines can range from 50 kW to several hundred megawatts for large utility power plants [64]. Because of the low partial load electric efficiency, steam turbines are not suitable for small-scale power plants. In the US and some European countries, steam turbines have already been widely installed in large-scale CHP/CCHP systems. If well maintained, the life of a steam turbine can be extremely long, and can be counted in years.

The working principle of the steam turbine is different from those of reciprocating IC engines and combustion turbines. For the latter two, electricity is the product and the heat is generated as a by-product. However, for the steam turbine, electricity is generated as the by-product. When equipped with a boiler, the steam turbine can operate with various fuels including clean fuels, such as natural gas, and other fossil fuels. This dramatically improves the flexibility of the steam turbine. In CHP/CCHP applications, the low pressure steam can be directly used for space heating or for driving thermally activated facilities.

GHG emissions of the steam turbine depends on the fuel it uses. If using some clean fuel, that is, natural gas, and adopting some effective emission control approaches, GHG emissions can be relative low. However, the low electric efficiency and long start-up time restrict the installation of steam turbines in small-scale CCHP systems and distributed energy applications [10]. Thus, steam turbines are only considered for being utilized in large-scale industries.

## 1.2.4   Micro-turbines

Micro-turbines are extensions of combustion turbines. A micro-turbine manufactured by Capstone Turbine Corporation is shown in Figure 1.2. The size of micro-turbines ranges from several kilowatts to hundreds of kilowatts. They can operate on various fuels, for example, natural gas, gasoline, diesel, and so on. One important characteristic of the micro-turbine is that it can provide an extremely high rotation speed, which can be used to efficiently drive the electricity generator. Because of the small size, micro-turbines are suitable for distributed energy systems, especially for CHP and CCHP systems. In the CCHP system, the by-product heat of the micro-turbine is used for driving sorption chillers and desiccant dehumidification equipment in summer, and to provide space heating in winter. The designed life of micro-turbines ranges from 40 000 to 80 000 hours [65, 66].

Another key advantage of the micro-turbine is the low level of GHG emissions; this is due to the *gaseous fuels feature lean premixed combustor* technology. In addition, low inlet temperature and high fuel-to-air ratios also contribute to emissions of $NO_x$ of less than 10 ppm. According to the data in [65], despite the stringent standard of less than 4–5 ppmvd of $NO_x$, almost all of the example commercial units have been certified to meet it.

Even with the drawbacks of higher capital costs than reciprocating engines, low electrical efficiency, and sensitivity of efficiency to changes in ambient conditions, the compact size and low weight per unit power, a smaller number of moving parts, lower noise, multi-fuel capability [67] and low GHG emissions still make the micro-turbine a popular prime mover in distributed energy systems. Analyses of CHP/CCHP systems installing micro gas turbines can be found in [67, 68]. The former discusses the

**Figure 1.2**    Capstone C200 micro-turbine with power output of 190 kW

potential of using micro-turbines in CCHP systems in distributed power generation. If the high capital cost and low efficiency can be solved, the market potential could increase dramatically.

In distributed energy systems, small-scale CCHP systems have been proven to be efficient. Due to the advantages, installing a micro-turbine becomes the best choice for a small-scale CCHP system. Much work has been done to investigate the performance of using micro-turbines in CCHP systems. Tassou *et al.* [69] validate the feasibility of the application of a micro-turbine-based trigeneration system in a supermarket. Beyond the feasibility, this paper also reveals that the economic viability of the system equipped with micro-turbines depends on the relative cost of natural gas and electricity. Karellas *et al.* [70] propose an innovative biomass process and use it to drive a micro-turbine and a fuel cell in a CHP system. The system efficiency can be extremely high when the gasification of biomass happens at high temperature. The innovative concept in this paper can be utilized in Biocellus. In 2002, the Oak Ridge National Laboratory (ORNL) presented its work of testing a micro-turbined CCHP system. The testing facility consists of a 30 kW micro-turbine for a distributed energy resource, whose exhaust is used to feed thermally activated facilities, including an indirect-fired desiccant dehumidifier and a 10-t indirect-fired single-effect absorption chiller [71]. From the test data, the efficiency of the micro-turbine strongly depends on the output level and ambient temperature, which makes the full power output to be preferred. Bruno *et al.* [72] conduct a case study of a sewage treatment plant, which is a trigeneration system. The prime mover selected in this system is a biogas-fired micro gas turbine. Hwang [73] in his work investigates potential energy benefits of a CCHP system with a micro-turbine installed. This paper gives the options to choose different types of chillers according to different configurations. Velumani *et al.* [74] propose and mathematically model a CCHP system with integration of a solid oxide fuel cell (SOFC) and a micro-turbine installed. This plant uses natural gas as the primary fuel and the SOFC is fed with gas fuel. Other evaluations, analyses and control strategy designs for CCHP systems running with micro-turbines can be found in [75, 76, 77, 78, 79], to name a few.

## 1.2.5    Stirling Engines

In contrast to the IC engine, the stirling engine is an external combustion engine, which is based on a closed cycle, where the working fluid is alternatively compressed in a cold cylinder volume and expanded in a hot cylinder volume [80]. Two basic categories of stirling engines exist: kinematic stirling engines and free-piston stirling engines. Also, the engine can fall into three configurations: alpha type, beta type, and gamma type [81, 82].

A stirling engine can operate on almost any fuel, for example, gasoline, natural gas and solar energy. Compared with IC engines, stirling engines operate with a continuous and controlled combustion process, which results in lower GHG emissions and less pollution [82, 83]. According to the data in [84], implementing a same capacity of 25 MW, the $NO_x$ emission of the stirling engine is 0.63 kg/MWh, compared with 0.99 kg/MWh of the IC engine. It is worth mentioning that, since the working fluid is sealed inside the engine, there is no need to install valves or other mechanisms, which makes the stirling engine simpler than an IC engine. As a result, stirling engines can be relatively safe and silent when running.

However, some challenges arise when using stirling engines in CHP/CCHP systems. The first one is the low specific power output compared with an IC engine of the same size. High capital cost is also a key factor that restricts its development. Another aspect is the working environment in CHP/CCHP systems. Unlike IC engines, the efficiency of a stirling engine drops when the working temperature increases. The last but equally important one is that the power output of the stirling engine is not easy to tune. Despite the above drawbacks, stirling engines have been installed in some CHP/CCHP applications because of the flexibility in fuel source, long service time and low level of emissions. A small-scale CHP plant with a 35 kW hermetic four cylinder stirling engine for biomass fuels has been designed, created, and tested by the Technical University of Denmark, MAWERA Holzfeuerungsanlagen GesmbH and BIOS BIOENERGIESYSTEME GmbH in Austria [85]. Moreover, SIEMENS collaborated with some European boiler manufactures, such as Remeha and Baxi, to conduct a large field test in 2009 and market introduction in 2010 of micro-CHP systems with stirling engines.

The most promising aspect of the stirling engine that it can be solar driven. Because of the increasing rate of carbon tax and more attention being paid to GHG emissions, the use of solar energy in CHP/CCHP systems gives more opportunities for the stirling engine.

Some theoretical work has also been done to investigate stirling engines installed in CCHP systems. Kong and Huang [84] propose a trigeneration system with a stirling engine installed and claim that this system could save more than 33% primary energy compared with the conventional SP system. Aliabadi *et al.* [86] discuss the efficiency and GHG emissions of a stirling-engine-based residential micro-CHP system fueled by diesel and biodiesel. According to the market assessment, stirling engines have not been widely applied in the CCHP market. To be further used in CHP/CCHP systems, solutions to high capital cost, long warm up time, and short durability of certain parts should be found [87].

## 1.2.6   Fuel Cells

Another environmentally concerned type of prime mover is the fuel cell. Fuel cells convert chemical energy from a fuel into electricity through a chemical reaction with oxygen or other oxidizing agents, and produce water as a by-product [88, 89, 90]. Compared with other fossil-fuel-based prime movers, fuel cells use hydrogen and oxygen to generate electricity. Since water is the only by-product, fuel cells are considered to be the cleanest method of producing electricity. Because of the few moving parts contained, the fuel cell system has a higher reliability than the combustion turbine or the IC engine [91]. However, some obstacles still exist for the development and application of fuel cells. The production of the materials, that is, oxygen, consumes energy and produces emissions. There are several ways, for example, electrolysis of water and generation from natural gas, to produce hydrogen for fuel cells, however, none of them can avoid both high energy consumption and high emissions. In the current market, there are various types of fuel cell, including proton exchange membrane fuel cell (PEMFC), alkaline fuel cell (AFC), phosphoric acid fuel cell (PAFC), molten carbonate fuel cell (MCFC) and the previously mentioned SOFC.

In recent years, much work has been done to investigate fuel cells in CCHP systems. The most widely used choice is the SOFC. Tse *et al.* [92] investigate a trigeneration system, which is jointly driven by a SOFC and a gas turbine, for marine applications. The efficiency of the configuration with double-effect absorption chiller can achieve 43.2% compared with 12% for the conventional system. In [93], Kazempoor *et al.* develop a detailed SOFC model, and study and optimize different SOFC system configurations. They also assess the performance of a building integrated with a trigeneration system, which comprises a SOFC and a thermally driven chiller. In [94], an SOFC with the capacity of 215 kW is combined with a recovery cycle for the sake of simultaneously meeting cooling load, domestic hot water demand and electric load of a hotel with 4600 m² area. An economic comparison between the trigeneration and SP systems indicates that, due to the lower heating value of the fuel, a maximum efficiency of 83% for energy trigeneration and heat recovery cycle can be achieved. Verda and Quaglia [95] model a distributed power generation and a cogeneration system incorporated with the SOFC. The authors also compare three configurations for this system, based on different choices of refrigeration systems, that is, single-effect absorption chiller, double-effect absorption chiller, and vapor compression chiller, from both technical and economic points of view. Other work on the environmental, economical and energetic analyses of CCHP systems equipped with SOFCs can be found in [96, 97, 98, 99, 100, 101, 102], to name a few. In [103], the authors model the CCHP system with stationary fuel cell systems from thermodynamic and chemical engineering aspects; and optimize the operation for that. Margalef and Samuelson [104] compare two strategies of operating a CCHP system equipped with a magnesium-air fuel cell (MAFC). The first strategy is to blend the exhaust gas with the ambient air; while the other one is to use the exhaust gas to drive an absorption chiller. The result shows that the second strategy is preferred, for the overall estimated efficiency is as high as 71.7%. Bizzarri [105] discusses the size effect of a PAFC system incorporated into a trigeneration system. Investigations reveal that the more the proper sizing is carried out for the highest environmental and energy benefits, the higher the financial returns will be. According to industry analysts Delta-ee, fuel cell CHP units have 64% of the CHP unit sale market, which doubles the results in 2011. It is becoming the most common technology employed in micro-CHP systems. In 2009, Tokyo Gas had success with the first commercialized residential fuel cell CHP systems based on PEFC, that is, ENE-FARM. In April 2013, they released the new version of that unit with lower cost, smaller size and lower $CO_2$ emission. This technology helps the fuel cell CHP take the market from other CHP prime movers. Ceramic Fuel Cell Limited's 1.5 kW SOFC CHP system, the so-called BlueGen, targets the market of social housing, shared accommodation, school and small business. This system can be beneficial for academic use, for all the data in this system can be pulled out for analyzing. However, feed-in tariffs and financial support are still obstacles for the development of this type of CHP system. California's Self Generation Incentive Program is a good example of incentive design for all policy makers.

The comparisons among different prime movers can be found in Table 1.1.

## 1.3   Thermally Activated Technologies

The most efficient solution to providing cooling is to utilize the rejected heat instead of electricity. This solution is realized by the thermally activated technology, which is dominated by the sorption cooling. The difference between the sorption cooling and the conventional refrigeration is that the former uses the absorption and adsorption processes to generate thermal compression rather than mechanical compression. One important reason for the CCHP system being efficient and with low GHG emissions is because space cooling and heating can be provided by using the rejected heat from the prime mover along with the electricity generation. This cascade utilization of heat is due to the thermally activated technology. In conventional SP systems, approximately two-thirds of the fuel used to generate electricity is wasted in the form of rejected

**Table 1.1** Comparisons among different prime movers

| Prime mover | Size (kW) | Pros | Cons | Emissions | Preferences and applications |
|---|---|---|---|---|---|
| IC engine | 10–5000 | Low capital cost<br>Quick start<br>Good load following<br>High partial efficiency<br>High reliability | Regular maintenance required | High $NO_x$ using diesel<br>Natural gas preferred | Working with absorption/electric chiller<br>Small- to medium-scale |
| Combustion turbine | 500–250 000 | High quality exhaust heat | Unacceptable low partial efficiency | $NO_x$ 25 ppm<br>CO 10–50 ppm | Applications with huge amount of thermal need<br>Large-scale |
| Steam turbine | 50–500 000 | Flexible fuel | Low electric efficiency<br>Long start-up | Depends on fuel | Electricity as by-product, thermal need preferred<br>Large-scale |
| Micro-turbine | 1–1000 | Flexible fuel<br>High rotation speed<br>Compact size<br>Less moving parts<br>Lower noise | High capital cost<br>Low electric efficiency<br>Efficiency sensitive to ambient conditions | $NO_x$ <10 ppm | Distributed energy system<br>Micro- to small-scale |
| Stirling engine | Up to 100 | Safer and silent<br>Flexible fuel<br>Long service time<br>Can be solar driven | High capital cost<br>Power output hard to tune | Less than IC engine | Solar driven<br>Small-scale |
| Fuel cell | 0.5–1200 | Operate quietly<br>Higher reliability than IC and combustion engine<br>High efficiency | Energy consumption and GHG emissions due to hydrogen producing | Extremely low | Micro- to medium-scale |

heat. By introducing thermally activated technologies, the electric load for cooling is shifted to the thermal load, which can be fully or partially achieved by absorbing or adsorbing the discarded heat from the prime mover. The main application of the sorption refrigeration is for CCHP systems in residential buildings, hospitals, supermarkets, office buildings, and district cooling systems [106].

Mainly, three types of thermally activated technologies exist, that is, absorption chiller, adsorption chiller, and desiccant dehumidifier. Since the temperature of the discarded heat from prime movers can lie in different ranges, thermally activated facilities should be chosen to couple with prime movers. For example, if the heat source temperature is around 540°C, then the suitable choice is a double-effect/triple-effect absorption chiller.

## 1.3.1   Absorption Chillers

Investigations of the absorption cycle began in the 1700s when it was found that ice could be produced by the evaporation of pure water from a vessel contained within an evacuated container in the presence of sulfuric acid [107]. The absorption chiller is one of the most commonly used and commercialized thermally activated technologies in CCHP systems. The difference between an absorption chiller and a vapor compression chiller is the process of compression. Since absorption chillers use heat to compress the refringent vapor instead of mechanically using rotating devices, they can be driven by the steam, hot water or high temperature exhaust gas. As a result, electricity needed for conventional refrigeration can be dramatically reduced, and the noise of the cooling process can be lowered significantly.

The working process of an absorption chiller can be divided into two processes: the absorption process and the separation process. The absorption process is shown in Figure 1.3. The left vessel contains the refrigerant and the right vessel is filled with a mixture of refrigerant and adsorbent. The absorption process of the refrigerant vapor

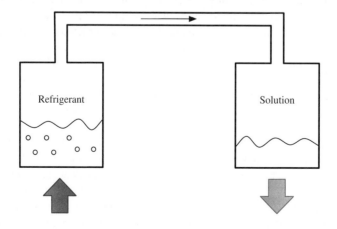

**Figure 1.3**   Absorption process

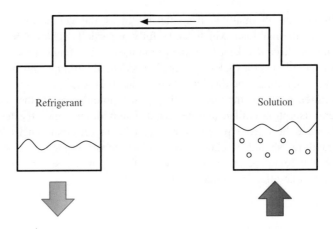

**Figure 1.4**   Separation process

in the right vessel will cause the pressure and temperature in the left vessel to drop. The temperature reduction in the left vessel is the refrigeration process. At the same time, as a result of the absorption process in the right vessel, heat must be rejected to the surroundings.

As the absorption process continues, the solution in the right vessel gradually becomes saturated. To reatin the ability to absorb, refrigerant must be separated from the solution. Figure 1.4 shows the separation process, which can be regarded as a reverse of the absorption process. Heat from the heat source is used to dry the refrigerant from the saturated or almost saturated solution. The refrigerant vapor is then condensed by a heat exchanger to act in the next cycle of the absorption process.

The chemical and thermodynamic properties of the working fluid determine the performance of an absorption chiller. The working fluid should be chemically stable, non-toxic, and non-explosive. Moreover, in liquid phase, it must have a margin of miscibility within the operating temperature range of the cycle [108]. According to [109], there are around 40 refrigerant compounds and 200 absorbent compounds available for the absorption chillers working fluid. However, the two most commonly used are lithium bromide/water (LiBr/water) and ammonia/water ($NH_3$/water). Usually, LiBr/water absorption chillers are used in air cooling applications with evaporation temperature in the range of 5–10°C, while $NH_3$/water absorption chillers are used in small-scale air conditioning and large industrial applications with evaporation temperature below 0°C [106].

In the literature, absorption chillers have been widely installed in CCHP systems. In 2002, the US government awarded Burns & McDonnell Engineering Co. a development contract of building an integrated gas turbine energy system based on improved CHP/CCHP technology. This plant is powered by a 4.6 MW Solar Turbines Centaur 50 gas turbine and two-stage indirect fired Broad Co. absorption chillers [110]. A small-scale CCHP system, installing a micro-turbine and an absorption chiller is demonstrated at the University of Maryland [111]. By adding the absorption chiller in the decentralized SOFC-based CCHP system, with an increased cost of about 0.7% compared with conventional systems, the $CO_2$ emissions can be reduced by 30%. In [102], a decentralized system with the integration of an SOFC and a double-effect LiBr/water absorption chiller is investigated. In [112], the authors introduce a CCHP system, with three engines and a total electrical power production of 9 MW, which supplies the thermal energy to drive an $NH_3$/water absorption chiller ARP-M10 by Colibri. This configuration is applied in a margarine factory in the Netherlands, a vegetable freezing factory in Spain, and a dairy factory in Spain. All of the three applications show a cost reduction by using this system.

### 1.3.2 Adsorption Chillers

The development of adsorption cooling began when the phenomenon of adsorption refrigeration caused by $NH_3$ adsorption on AgCl was discovered by Faraday in 1848 [113]. Similar to absorption chillers, adsorption chillers make use of the discarded heat from the prime mover to provide space air conditioning. One important difference of an adsorption chiller from an absorption chiller is that the former can be driven by a low temperature heat source. Furthermore, the noiselessness, being solution pump free, the lack of corrosion and crystallization, and small volume make adsorption chillers suitable for CCHP systems, especially small-scale ones [114, 115, 116].

Different from the absorption chiller, in which a fluid permeates or is dissolved by a liquid or solid, the adsorption chiller provides cooling by using solid adsorbent beds to adsorb and desorb a refrigerant. Similar to the two processes in the absorption chiller, temperature of the adsorbent changes according to the refrigerant vapor adsorbed and desorbed by adsorbent beds. A simple adsorption refrigeration circuit consists of a solid adsorbent bed, a condenser, an expansion valve, and an evaporator [34]. The refrigeration process of the adsorption chiller can also be divided into two processes, that is, absorbent heating and desorption process, and the adsorption process. In the first process, the adsorbent bed is connected with a condenser first. Driven by a low temperature heat source, the refrigerant is condensed in the condenser and heat is released to the surroundings. Following that, in the adsorption process, the adsorbent bed is connected to an evaporator, and at the same time disconnected from the condenser. Then cooling is generated from evaporation and adsorption processes of the refrigerant. However, this simple adsorption chiller provides cooling in an intermittent way. To continuously provide cooling, two adsorbent beds should be installed in the system together, in which one bed is heated during the desorption process and the other one is cooled during the adsorption process.

The same as in absorption chillers, adsorption chillers have no internal mechanical moving parts. As a result, they not only run quietly, but also need no lubrication and less maintenance. In addition, adsorption chillers are always made in modules, which makes them suitable for the cooling capacity expansion. Moreover, as mentioned before, since no electricity and fuel is needed to drive the chiller, a low level of GHG emissions is guaranteed. Because of the advantages of the adsorption chiller, research and demonstrations of this type of chiller installed in the CCHP system have been developed widely. Li and Wu [117] discuss the performance of a silca gel/water adsorption chiller in a micro-CCHP system according to different working conditions, especially for different electric loads. In 2000, a CCHP system, equipped with a fuel cell, a solar collector and an integration of a mechanical compression chiller and an adsorption chiller, was installed in the St. Johannes Hospital, Germany [10]. In the same year, a CCHP system with an adsorption chiller began to operate in the Malteser's Hospital, Germany. Shanghai Jiao Tong University (SJTU) has been investigating the applications of adsorption chillers in CCHP systems for many years. In 2004, SJTU set up a gas-fired micro-CCHP system consisting of a small-scale power generator set and a novel silica gel/water adsorption chiller [106].

## 1.3.3   Desiccant Dehumidifier

A desiccant dehumidifier removes the humidity from the air by using materials that attract and hold moisture. To achieve comfort cooling, sensible cooling, aiming to lower the air temperature, and latent cooling, which means reducing humidity, should be achieved simultaneously. Since, by introducing desiccant dehumidifiers, the control of humidity is independent of the temperature, potentially wasted thermal energy can be used to reduce the latent cooling load and bacteria and viruses can be scrubbed out; desiccant dehumidifiers always operate with chillers or conventional air-conditioning systems to provide comfort cooling and to increase overall system efficiency [10, 118].

Mainly, there exist two commercialized types of desiccant dehumidifiers, which are distinguished by desiccant types, that is, solid desiccant dehumidifier and liquid desiccant dehumidifier. Solid desiccant dehumidifiers are usually used for dehumidifying air for commercial HVAC systems, while liquid desiccant dehumidifiers are popular in industrial or residential applications. Desiccant dehumidifiers are suitable for CHP/CCHP systems, because the regeneration process in the desiccant system provides an excellent use of waste heat [119]. In [120], the authors introduce a CCHP system utilizing the solid desiccant cooling technology. Researchers in Tsinghua University, China, also carried out laboratory research to assess the operational performance and energy efficiency of a CCHP system installing a liquid desiccant dehumidifier [121]. The data collected from summer and winter show that the only way to increase the overall efficiency is to install more waste heat driven equipment to utilize the low-quality waste heat. In [52], the authors assess a desiccant dehumidifier system in a CHP application incorporating an IC engine. Badami and Portoraro [122] analyze the performance of a trigeneration plant with a liquid desiccant cooling system installed at the Politecnico di Torino, Italy. In this paper, the authors provide both the energetic and economic analyses to show the huge potential of using this type of trigeneration system. The desiccant dehumidifier system allows the temperature and humidity in the classroom to be controlled. The air conditioning service can also make this system suitable for academic use. It is proved that with the liquid desiccant cooling system, we can make full use of the waste heat to provide a cooling service in the summer and the overall efficiency can be dramatically increased.

The comparisons among different thermally activated technologies can be found in Table 1.2.

## 1.4   System Configuration

An economical, efficient and of low emissions CCHP system should be designed with full consideration of energy demands in a specific area, prime mover and other facilities' types and capacities, power flow and operation strategy, and the level of GHG emissions. The selection of facility types belongs to the design of the system configuration, which emphasizes the selection of prime movers according to current available technologies, and on the system scale. It is well known that different climate conditions in different areas lead to different patterns of energy demands. For example, in conventional CHP systems, steam-turbine-based plants are always used as heat plants with electricity generated as a by-product in some cold areas. While in the temperate zone, in summer, the amount of electricity needed by the air conditioning could be huge. Thus, in this kind of area, combustion-turbine-based CHP systems are popular. Some CHP/CCHP applications based on prime mover selections have been mentioned in Section 1.2. The existing CHP/CCHP sites in the market sorted by prime movers are shown in Figure 1.5. With a selected CCHP system configuration, operation strategy is the key to achieve the most efficient way for the CCHP to operate. The operation strategy determines how much electricity or fuel should be input to the system according to the demands; which facility should be shut down to keep the whole system efficient; how the energy carriers flow between facilities; and how much power one facility should operate at. With a designated configuration and an appropriate operation strategy, suitable sizing and optimization can make the system operate in an optimal way. Here, CCHP applications categorized by plant size will be mainly discussed.

Categorized by the rated electricity generation capacity, the CCHP system can have micro-scale (under 20 kW), small-scale (20 kW–1 MW), medium-scale (1 MW–10 MW), and large-scale (above 10 MW).

### 1.4.1   Micro-Scale CCHP Systems

Micro-scale CCHP systems are the ones with rated size under 20 kW. Recently, much work has been done to investigate and analyze micro-scale CCHP systems, for they are suitable for distributed energy systems. In the literature, Easow and Muley [123] discuss the potential of the micro trigeneration system being applied in the decentralized cooling, heating, and power. By testing an experimental plant, which is a micro trigeneration system with a liquefied petroleum gas driven Bajaj 4-stroke IC engine, the increased energy supply reliability and security, lower energy cost, higher efficiency and less fuel energy loss are verified in this paper. In 2010 in the North Carolina Solar Center, Raleigh, North Carolina, an integrated micro-CCHP and solar

**Table 1.2** Comparisons among different thermally activated technologies

| Facility | Capacity | Pros | Cons | COP | Preference |
|---|---|---|---|---|---|
| Absorption chiller | 10 kW–1 MW | Driven by steam <br> Low noise <br> Can be driven by low-quality heat source <br> Low GHG emission | Less efficient than compressor-drive chiller | Up to 1.2 | LiBr/water: evaporation temperature 5–10°C <br> NH$_3$/water: evaporation temperature <0°C <br> Small- to large-scale <br> Double-effect preferred |
| Adsorption chiller | 5.5–500 kW | Driven by steam <br> Small size <br> Noise free <br> Corrosion and crystallization trouble free <br> No lubrication <br> Low GHG emissions | Can only be driven by high quality heat <br> High capital cost | 0.6 | Small-scale |
| Desiccant dehumidifier | N/A | Control of humidity independent of the temperature is allowed <br> Reduce the mechanical cooling load | High capital cost <br> Regular maintenance required | N/A | Solid: HVAC systems <br> Liquid: Industrial and residential applications |

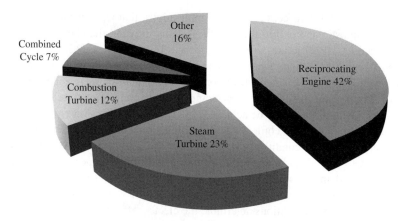

**Figure 1.5** Existing CHP/CCHP sites classified by prime movers

system was installed to demonstrate the technical and economic feasibility of incorporating photovoltaic (PV), solar thermal, and propane-fired CHP systems into an integrated distributed generation system [124]. The rated output of the CHP plant, PV, and solar thermal system is 4.7 kW plus 13.8 kW, 5.4 kW, and 4.1 kW, respectively. Thermal energy produced by this system can be used for space heating, domestic hot water, process heating, dehumidification and absorption cooling. With this solar-based CCHP system, CO and $NO_x$ emissions can be reduced to below 250 ppm and below 30 ppm, respectively. Other experimental and test results of micro-CCHP systems can be found in [125, 126, 127], to name a few. In addition, various work on energetic, economic and thermodynamic analyses has been done in recent years. In [128], the authors provide an analysis of matching prime mover heat sources to thermally driven devices in a micro-scale trigeneration system. The T–Q analysis of the prime mover waste heat in this literature also indicates the promise of incorporating micro-turbines, SOFCs and HT-PEMFCs into the trigeneration system. Other analyses can be referred to [129, 130, 131, 132, 133, 134]. Some researchers also focus on the optimization of micro-CCHP systems. Arosio *et al.* [135] model a micro-scale CCHP system based on the linear optimization and incorporate the Italian tariff policy into this model. The proposed model allows to evaluate the influence of each parameter on the system's performance. Other optimization research can be found in [127, 136]. Recently, some renewable energy, such as solar energy, has been implemented in the CCHP system to further reduce GHG emissions. In [137], a micro trigeneration system equipped with a solar system is studied. This system is integrated by a micro-turbine with output power of 5 kW and a LiBr/water absorption chiller. The heat source for the absorption chiller and the micro CHP system is a solar storage tank. Installing the solar system can efficiently increase the overall efficiency and consume less primary energy. Immovilli *et al.* [138] compare a conventional CCHP system with one based on solar energy. The configuration of a PV collector coupled with a vapor compression cooling system is verified to be the best commercially available solution. Besides the PV, they also propose two technical solutions for the solar CCHP to access residential applications, that is, concentrated sunlight all-thermoacoustic and hybrid thermo-PV systems. Beyond the above, some novel micro-scale CCHP structures are also proposed. Henning *et al.* [120] investigate a micro trigeneration system, whose air conditioning facilities integrate a vapor compression chiller and a desiccant wheel, for indoor air conditioning in a mediterranean climate. The research results show that, compared with conventional technologies, an electricity saving of 30% can be achieved. In [139], the authors investigate the performance of an absorption chiller, which is installed in a micro-scale BCHP system, under varying heating conditions. Huangfu *et al.* [5] introduce a novel micro-scale CCHP system which can be applied in domestic and light commercial applications. Evaluations and analyses in the literature show that this micro CCHP system enjoys good economic efficiency with a payback period of 2.97 years, which is quite short; also the electric load conditions determine the electric efficiency, which means that, compared with half load, the system can perform better when operating at full load.

## 1.4.2 Small-Scale CCHP Systems

Small-scale CCHP systems are the ones with rated size ranging between 20 kW and 1 MW. They have been widely used in supermarkets, retail stores, hospitals, office buildings, and university campuses. Different types of prime movers and refrigeration systems can be combined freely according to energy demands. Around the world, small-scale CCHP plants have been installed in many applications. A 500 kW biomass CCHP plant is installed in the Cooley Dickinson Hospital, Northampton, Massachusetts, which is a 55 742 m$^2$ hospital with 140 beds [140]. In 1984, the first boiler installed in this system was a Zurn-550 HP biomass boiler, which was fired by virgin wood chips. Then in 2006 and 2009, due to increased energy demands, an AFS-600 HP water/fire tube high pressure boiler, two 250 kW Carrier Energent micro steam turbines, and a 2391 kW absorption chiller were installed consecutively. This CCHP system has brought a lot of benefits to this hospital, especially the 99.5% particulate removal accomplished by the Multiclone separator and Baghouse. The East Bay Municipal Utility District (EBMUD), which was a publicly owned utility that provided a water service to portions of two counties in the San Francisco bay area, began to use a 600 kW micro-turbine CHP/chiller system at its downtown Oakland administration building in 2003 [141]. This system is composed of ten 60 kW Capstone micro-turbines and one 633 kW YORK absorption chiller. The total project cost $2 510 000, with a payback period estimated to be 6–8 years. Another small-scale CCHP application is in the Smithfield Gardens, which is a 56-unit affordable assisted-living facility in Seymour, Connecticut [142]. This system includes a 75 kW Aegen 75LE CHP module, an American Yazaki absorption chiller, and a Baltimore Air Coil cooling tower. With this system installed, the Smithfield Gardens can save 22% on its annual energy costs. With the pollution controlled by the Non-Selective 3-way Catalytic Reduction System, $CO_2$ and $NO_x$ reductions can reach 32% and 74%, respectively. Vineyard 29, a winery located in St. Helena, California, installed a 120 kW micro-turbine/chiller system to reduce GHG emissions as well as toxins into the environment [143]. Two 60 kW Capstone C60 micro-turbine systems were installed to provide electricity and thermal energy. This a good example of waste heat cascade utilization. Through the heat recovery system, hot water is produced for the wine processing, and the other part of the thermal energy is adsorbed by a 70 kW Nishiyodo adsorption chiller to provide space cooling. In addition, a Dolphine pulsed power system is used in the EvapCo cooling tower. With a total capital cost of $210 000, the estimated payback period is 6–8 years, which means $25 000–38 000 per year. In [144], the authors provide experimental results for a real small-scale CCHP system operating at full load and partial load. This test plant, consisting of a 100 kW natural gas-powered micro-turbine and a liquid desiccant system, is installed at the Politecnico di Torino, Turin, Italy. Comparisons of primary energy savings (PES) among different prime mover load situations are made. The data shows that adopting the partial load strategy can cause an energetic performance decrease. Katsigiannis and Papadopoulos [145] conduct two case studies in the indoor Swimming Pool Building, and the Law School Building in the Democritos University of Thrace, Greece, to investigate their systematic computational procedure for assessing a small-scale trigeneration system. The procedure includes an indirect estimation of pertinent loads, indoor swimming pool heating, CCHP facility selection, system sizing, and economic evaluation. This work further verifies that the system performance mainly depends on the system size. Less cost and pollution can be readily observed from the simulation. Other applications and test work can be found in [146, 147, 148].

In their theoretical work, Chicco and Mancarella [149] summarize some key issues and challenges of the planning and design problems for a small-scale trigeneration system. Time domain simulations are conducted to assess each energy vector production within the system by introducing new performance indicators, that is, trigeneration primary energy saving (TPES), electrical-side incremental trigeneration heat rate (EITHR), thermal-side incremental trigeneration heat rate (TITHR) and cooling-side incremental trigeneration heat rate (CITHR). Other energetic, economic and thermodynamic analyses can be referred to [122, 150, 151]. In [122], the authors investigate an innovative natural-gas-based CCHP system, whose electricity, heating and cooling capacities are 126 kW, 220 kW and 210 kW, respectively. The gas-fired IC engine works in pairs with a liquid LiCl/water desiccant cooling system. The authors also give energetic and economic analyses, including the influence of the fuel and electric price, and index variations due to the plant cost, of this system. The proposed system has a payback period of around 7 years, and will provide $200 000–220 000 net present value after 15 years.

Some researchers focus on optimization problems involved in the small-scale CCHP system design. Abdollahi and Meratizaman [152] propose a multi-objective optimization method for a small-scale distributed CCHP system design. The environmental impact objective function is defined to be the cost. An economic analysis is conducted using the total revenue requirement (TRR) method. They adopt the genetic algorithm (GA) to find a set of Pareto optimal solutions; and apply the risk analysis to complete the decision-making to find the optimal solution from the obtained set. In [153], an optimization problem of the energy management in the CCHP system is solved by mixed integer linear programming (MILP). The solution aims to control the on/off status of system components. A comparison, with data collected from a 985 kW plant, is made between the proposed energy management and conventional management. The proposed optimal strategy allows a 1 year reduction of the payback period. Hossain *et al.* [154] present a design and the construction of a novel small-scale trigeneration system driven by neat non-edible plant oils, including jatropha and jojoba oil. The use of local available non-edible oil means this plant can run without depending on imported petroleum fuel, which results in a high economical efficiency. Moreover, GHG emissions can be dramatically reduced by using the rejected heat from the prime mover to provide cooling and heating.

## 1.4.3   Medium-Scale CCHP Systems

As mentioned, medium-scale CCHP systems are those with rated power ranges of 1–10 MW. From this level of rated size, CCHP systems begin to operate in large factories, hospitals, schools, and so on. A 4.3 MW CCHP plant has been serving the Elgin Community College, Elgin, Illinois, since 1997 [155]. The first phase of this plant, which was a 3.2 MW CCHP plant, was installed in 1997 to provide electricity, low pressure steam, and absorption cooling to main campus buildings. In 2005, due to campus expansion, the generation set and the absorption chiller were both expanded in the second phase. The prime mover in this system is a combination of four 800 kW Waukesha reciprocating engines and one 900 kW Waukesha reciprocating engine. Cooling is provided by one YORK 1934 kW absorption chiller and one Trane 2813 kW absorption chiller. The heat recovery equipment includes five Beaird heat recovery silencers and five Beaird exhaust silencers. The two phases cost $2 500 000 and $1 200 000, respectively. The payback period for the second phase is about 4 years with annual savings of around $300 000. Another medium-scale CCHP application is that with a capacity of 3.2 MW which was installed in Mountain Home VA Medical Center, Mountain Home, Tennessee, in 2011 [156]. This medical center serves 170 000 military veterans in the surrounding states. The whole plant consists of one 3.2 MW dual-fuel engine generator set, fired by landfill bio-gas, two 1.8 MW back up diesel-fired engine generator sets, a heat recovery steam generator (HRSG), and a 3.5 MW absorption chiller. With this plant, the estimated cost savings over 35 years can be $5–15 million. In the University of Florida, a 4.3 MW CHP plant began serving the Shands HealthCare Cancer Hospital in 2008 in order to solve the problem of increasing electricity and fuel prices, to reduce budget, and to reduce GHG emissions. The total installation cost $45 million. This system, designed by Gainsville Regional Utilities (GRU), consists of one 4.3 MW combustion turbine, one 6.5 t/h HRSG, one 4.2 MW steam turbine centrifugal chiller, two 5.3 MW electric centrifugal chillers, and one 13.6 t/h packaged boiler. The 4.3 MW natural gas turbine provides 100% of the hospital's electric and thermal needs. By using this system, a total thermal efficiency of 75% can be achieved.

Other applications of medium-scale CCHP systems can be found in [157, 158, 159], to name a few.

Moreover, researchers are also concerned with theoretical research on medium-scale CHP/CCHP systems. In [160], in order to raise the energy efficiency, the authors propose a trigeneration scheme for a natural gas processing plant by installing a turbine exhaust gas waste heat utilization. This trigeneration system makes use of the rejected heat from the gas turbine to generate process steam in a waste HRSG. A double-effect LiBr/water absorption chiller is driven by the process steam to provide space cooling; another part of the process steam is used to meet furnace heating load and to supply plant electricity in a combined regenerative Rankine cycle. The measured CCHP power output is 7.9 MW. The expected annual operating cost savings can reach as high as $20.9 million with only 1 year payback period. In [161], the authors design a CCHP system for a business building in Madrid, Spain. The basic demands of this building are 1.7 MW of electricity, 1.3 MW of heating and 2 MW of cooling. By designing the operation strategy and optimizing facility capacities, the final design of the configuration is given to be an integration of three 730 kW IC engines, one 3 MW double-effect absorption chiller, one conventional chiller of 4 MW, and one 200 kW boiler for back up. On account of incorporating a thermal solar plant, the capital cost is €3.32 million, which is expected to be paid back in 11.6 years. Compared with conventional trigeneration systems, PESs in this plant increase a lot due to the incorporation of the thermal solar plant into the trigeneration plant. In [162, 163], the authors propose a methodology for thermodynamic and thermoeconomic analyses of a trigeneration system equipped with a Wartsila 18V32GD model 6.5 MW gas-diesel engine. This system is installed in the Eskisehir Industry Estate Zone, Turkey. Efficiencies of energy, exergy, Public Utility Regulatory Policies Act (PURPA), and equivalent electrical of the trigeneration system are determined to be 58.94%, 36.13%, 45.7%, and 48.53%, respectively. This CCHP system can also be transplanted to an airport to provide cooling, heating, and electricity. Other theoretical work can be referred to [164, 165].

### 1.4.4 Large-Scale CCHP Systems

Large-scale CCHP systems are categorized as those with output power of above 10 MW. This type of CCHP system can provide substantial electricity for industry use, and vast heating and cooling for universities and residential districts, which have a high population density. So far, to combat the problem of GHG emissions and increased prices of electricity and fuel, an increasing number of large-scale CCHP systems have been installed. The University of Michigan, Ann Arbor, Michigan, began to adopt the cogeneration system in 1914 [166]. Combined with absorption chillers, this 45.2 MW CHP plant consists of: six conventional gas-/oil-fired boilers from Combustion Engineering, Wickes, Murray and Foster Wheeler (a total of 453.6 t/h of steam capacity); three Worthington back-pressure/extraction steam turbine generators (rated at 12.5 MW each); two gas/oil solar combustion turbines (3.7

and 4 MW, respectively); and two Zurn HRSGs with supplemental gas firing (29.5 t/h each). The electricity production of this plant can rarely reach the maximum capacity, for the plant has to provide steam for other use, such as, in summer, the absorption chiller. The system installed saved the university $5.3 million in 2004. In San Diego, California, the University of California at San Diego installed a 30 MW polygeneration plant in 2001 [167]. The 30 MW combined cycle is composed of two 13.5 MW Solar Turbines Titan 130 gas turbine gen-sets and a 3 MW Dresser-Rand steam turbine. The rejected heat is used to run a steam driven centrifugal chiller; to provide domestic hot water for campus use; and to run the steam turbine for additional electricity production. The whole system can achieve 70% gross thermal efficiency. Annually, by installing this system, $8–10 million can be saved. In this site, an emission control system, that is, SoLoNOx$^{TM}$, is adopted to control the level of NO$_x$ emissions to 1.2 ppm, which is much lower than the permitted 2.5 ppm. Another classic large-scale CCHP application is the plant installed in the University of Illinois at Chicago [168]. This plant, established from 1993 to 2002, is separated into two parts: the east campus system and the west campus system. In the east campus CCHP plant, two 6.3 MW Cooper–Bessemer dual-fuel reciprocating engine generators and two 3.8 MW gas reciprocating engine generators are installed as the prime mover. Cooling is provided by one 3.5 MW Trane two-stage absorption chiller, two 7 MW YORK electrical centrifugal chillers, and several remote building absorption chillers. The capital cost of $25.7 million is estimated to be paid back in 10 years. PESs, CO$_2$ reductions, NO$_x$ reductions, and SO$_2$ reductions can reach 14.2%, 28.5%, 52.8%, and 89.1%, respectively. In the west campus, because of the large energy demand in the hospital and several buildings, an additional 37.2 MW generation set, composed of three 5.4 MW Wärtsilä gas engines and three 7 MW Solar Taurus turbines, has been added. Besides the prime mover, an additional 7 MW absorption chiller was also installed in the west campus CCHP plant. With the capital cost of $36 million, the payback period is estimated to be 5.1 years. Other applications that show the success of using the CCHP scheme in large-scale systems can be found in [169, 170, 171].

The comparisons among different system configurations can be found in Table 1.3.

**Table 1.3** Comparisons among different system configurations

|  | Size | Preference |
| --- | --- | --- |
| Micro-scale | <20 kW | Distributed energy system |
| Small-scale | 20 kW–1 MW | Supermarkets, retail stores, hospitals, office buildings, and university campuses |
| Medium-scale | 1 MW–10 MW | Large factories, hospitals, and schools |
| Large-scale | >10 MW | Large industries<br>Waste heat can be used for universities and districts with a high population density |

## 1.5    System Management, Optimization, and Sizing

Once the configuration of a CCHP system is determined for a specific application, the next step is to manage the energy flow reasonably and to select an appropriate facility capacity to achieve maximum cost and emission reduction. Actually, in some recent system optimization work, the operation strategy, power flow and facility size are optimized simultaneously. In this Section, some conventional and novel operation strategies, system optimization approaches, and sizing work will be introduced.

### 1.5.1    Conventional Operation Strategies

Two classical operation strategies for the CHP/CCHP systems are following electric load (FEL) and following thermal load (FTL) [172, 173], which can also be referred to as electric demand management (EDM) and thermal demand management (TDM) [174]. In the FEL strategy, the CCHP system first purchases the fuel to provide enough electricity for the building users. If the excess heat cannot meet the cooling and heating demand, additional fuel should be purchased to feed the auxiliary boiler to generate enough thermal energy. In the FTL strategy, the CCHP system first meets the thermal demand, including the cooling and heating, then if the electricity provided by the PGU cannot meet the building users' demand, additional electricity should be purchased from the local grid to compensate for the deficit. However, both the FEL and FTL strategies can inherently waste a certain amount of energy. This is because, for instance, when the CCHP system runs under the FEL strategy to provide enough electricity for the building users, if the thermal demand is less than the thermal energy PGU provides, the excess thermal energy will be wasted. It is a similar case for the FTL strategy. The comparisons and analyses of the two strategies are investigated in [33, 174, 175, 176, 177, 178, 179, 180], to name a few.

### 1.5.2    Novel Operation Strategies

In order to reduce the energy waste and to reduce primary energy consumption (PEC), annual total cost (ATC), and GHG emissions, it is necessary to design an optimal operation strategy. Due to different definitions of "optimal", the operation strategies designed are different. In [181], Liu *et al.* propose a novel operation strategy for the CCHP system by using the concept of "balance". By adjusting the electric cooling to cool load ratio between the electric chiller and absorption chiller, the balance point of users, electric demands, cooling demands and heating demands evolves to a balance space. Optimal operation strategy is designed to keep the energy balance. The case study shows that, running under the proposed operation strategy, the PEC, carbon dioxide emissions (CDE), and the ATC are much lower than those of the SP system. In [182], the author proposes an optimal operation strategy for an offline non-linear model, that is, TOOCS-off, of a CCHP system. This optimization model considers the electric and thermal load in each time interval, prices of electricity sold to costumers or purchased from utility, and prices of heating and cooling. In the cost function of the TOOCS-off model, the total economic benefit of this system is maximized during total daily operation time. In the constraints, facilities' thresholds and output upper bounds are considered simultaneously. A CCHP system with a capacity of 143 kW, equipped with a 450 kW auxiliary boiler, a 600 kW absorption chiller, and a 800 kWh content heat storage tank, is used to verify the feasibility of this offline model and the optimal operation strategy. Based on source PEC, Fumo and Chamra [183] analyze four CCHP system operation conditions, including power and cooling without requiring boiler operation (in spring/autumn), power and cooling requiring boiler operation (in summer), power and heating without requiring boiler operation (in spring/autumn) and power and heating requiring boiler operation (in winter). The results of this study can contribute to the design of the operation strategy to reduce undesired increase of energy consumption. In [175], the authors design an optimal operation scheme for a CCHP system by considering the PEC and emissions of pollutants besides the energy cost. The operation is optimized by an optimal energy dispatch algorithm. The evaluation of the performance of a CCHP system, operating under the proposed strategy, is conducted using five cities' realistic climate data. Cardona *et al.* [184] propose a profit-oriented optimal operation strategy, considering both the articulated energy tariff system and the technical characteristics of components. In [185], instead of the profit-oriented strategy and the primary energy-oriented strategy, the authors adopt an emission-oriented strategy in order to reduce GHG emissions. The control scheme in the proposed strategy is an on–off control, that is, if the level of GHG emissions is greater than a specific value, then the PGU should stop; otherwise, the PGU runs to meet the energy demand. A comparison of GHG emissions of the proposed strategy, profit-oriented strategy and primary energy-oriented strategy is made to show the effectiveness of the proposed operation strategy. In [186], the authors propose an FEL/FTL switching operation strategy for the CCHP system. An integrated performance criterion, including PEC, CDE and operational cost (COST), is used to determine the switching action between FEL and FTL strategies. However, the inherent energy waste still exists. In [187], by considering the uncertainties of the price of the purchased electricity, the delivered demand for electricity, and the marginal cost of self generation, the authors propose an operation strategy design method using a risk management approach. By using the risk metrics, the steam and gas turbine generated electric power, the benefits and costs can be forecast. Moreover, an optimal control tool, that is, the model predictive control (MPC), is also used in [187] to schedule the operation strategy. Mago and Chamra [172] propose an optimized operation strategy, which can be referred to as following the hybrid electric-thermal load (HETS). The analyses and evaluations show that, when operating under the HETS, a CCHP system can perform better in the aspects of PEC, operational cost and CDE, when compared with FEL and FTL strategies. In [188], the optimization of the operation strategy is formulated to be a linear programming (LP) problem with the objective function set to be the operation variable cost. This problem is constrained by capacity limits, equipment efficiencies, energy balance equations, and demand constraints. The obtained optimal operation strategy is classified in nine operational modes due to the price of electricity from grid, electricity sold back price, auxiliary heat, and waste heat. A thermoeconomic analysis, based on the marginal cost, is also conducted to investigate the relationship between the optimal operation mode and energy demands, as well as the prices of consumed resources. Aiming to maintain the system autonomy to ensure the grid reliability and to minimize excess power production, Nosrat and Pearce [189] propose a dispatch strategy for a PV-CCHP system, in which the thermal energy waste can be significantly reduced. Decision-making of this dispatch strategy depends on the output of the PV array and is separated into four steps. In each step, several operation conditions are analyzed to choose the strategy between FEL and FTL. The results show that an improvement of 50% can be achieved by using this dispatch strategy. Because the strategy is chosen from FEL and FTL directly, the inherent energy waste still cannot be avoided.

### 1.5.3  System Optimization

To optimize the system performance, a mathematical model should be constructed first to make use of the various optimization algorithms. In the literature, much work on the optimization has been done to investigate the CCHP optimization problem. Among these approaches, due to the on–off character of the components, mixed integer programming is the most widely used. Based on the concept of *superstructure*, the authors in [190] propose a systematic method to optimize the size of a CCHP system powered by natural gas, solar energy, and gasified biomass. Modeled by the mixed integer non-linear programming (MINLP) model, PESs, GHG emissions, and economic feasibility are optimized. They also point out that the trade-off between the economical and environmental concerns should be taken into consideration when designing a CCHP system. Following the previous work, which only concerns the monthly average requirement, Rubio-Maya *et al.* [191] take the hourly data, analysis, and energy storage system into consideration. This NP problem is solved by a generalized-reduced-gradient (GRG)-based algorithm. In [192], Buoro *et al.* model a trigeneration system to be an MINLP model. They propose a scheme of several buildings connected with each other. Thus, the optimal solution of this problem contains the prime mover's type and size, positions of district heating and cooling (DHC) pipelines, and the operation of each system component. Besides considering the thermodynamics of each system component, the objective function also takes the facilities' cost and DHC network cost into consideration. Moreover, the influence of various amortization periods on the optimal solution is also discussed. Li *et al.* [193] model and optimize a system by an MINLP model. Analyses in this literature show that the optimal facility size and the economic performance of the whole system mainly depend on the average energy demand. In [194], MILP is used to model and optimize a CCHP system with a thermal storage system installed and to minimize the ATC. In this reference, the effect of legal constraints and different operation modes on the optimal design is also discussed. In [195], the authors construct an MINLP model for increasing the power production in a small-scale CHP plant. This CHP plant is driven by a steam Rankine process fired by biomass fuel. Using the MINLP, due to the complicated decision-making process in the system, the optimization problem is modeled to be a non-convex problem. This problem is solved several times to filter out suboptimal solutions and to find out the most likely global optimal solution. This model is also tested on four existing CHP plants, in which the result shows that, by adding a two-stage district heat exchanger, a preheater, a steam reheater and a fuel dryer, the electric efficiency and power to heat ratio can be increased. Arcuri *et al.* [38] carry out a mixed integer programming model for the optimization of a CCHP system in a hospital. The optimization results, including short-term optimization and long-term optimization, give the optimal plant design, that is, facility sizes and running conditions. With the proposed optimization approach, the case study result shows that, by utilizing size optimized heat pumps, the trigeneration system can be improved in terms of economic, energetic and environmental aspects. Li *et al.* [196] thermoeconomically optimize a distributed trigeneration system by considering thermodynamic, economic and GHG emissions aspects. This optimization includes the system configuration and operation strategy. With the objective function set to be the system net present value (SNPV), MINLP is used to model the system and the GA is adopted to solve it. An optimal solution is found under different economic and environmental legislation contexts in Beijing, China. Rong and Lahdelma [197] model a trigeneration system by the LP model with three components' characteristics. The objective function is set to be a combination of production and purchase cost, and the carbon cost. This problem is solved by the Tri-commodity algorithm, which is 36–58 times faster than an efficient Simplex code. Using the PGU capacity as the decision variable, Wang and Fang [198] optimize a CCHP system by the GA. The objective function is set to be a weighted summation of PEC, ATC, and CDE. Wang *et al.* [32] design an optimal operation strategy by optimizing the capacity of the PGU, the capacity of the heat storage tank, the on–off coefficient of the PGU and the ratio of electric cooling to cool load using the particle swarm algorithm. All of the four decision variables are globally optimized, that is, fixed once determined. The authors of this reference also compare the result of [32] to that of another reference [33]. In [33], only the PGU capacity and electric cooling to cool load ratio are considered to be decision variables. The objective function, which includes PESs, annual total cost savings (ATCS) and CDER, is minimized by the GA. When compared with the GA, the particle swarm algorithm converges faster and the result is better. Sheikhi *et al.* [199] conduct a cost–benefit analysis of a CHP system aiming to maximize the benefit-to-cost ratio. With the benefit-to-cost ratio incorporated into the objective function, using the concept of *energy hub*, the size and efficiency of this CHP system is optimized using the evolutionary-algorithmic (EA) approach. Kavvadias and Maroulis [200] set up a multi-objective optimization problem for a trigeneration system, in which facilities, sizes, pricing tariff schemes and the operation strategy are to be optimized according to realistic conditions. Pointing out the drawbacks of traditional load following strategies, the authors propose a new load following strategy, that is, electric/heat equivalent load follow, which includes the continuous operation, peak shaving, electricity equivalent demand following, and heat equivalent demand following. The optimization problem is solved by the multi-objective

EA approach. Wang *et al.* [201] analyze the energy consumption and construct an environmental impact model, consisting of the global warming, acid precipitation and stratospheric ozone depletion, for an SP system and a CCHP system. The system capacity is optimized by the GA following the FEL strategy. In [202], the authors model a trigeneration system using a fuzzy multi-criteria decision-making model. Different configurations of trigeneration systems are compared with an SP system under this model. This fuzzy multi-criteria model can help to choose the optimal trigeneration configuration from technical, economical and some external (like the environmental) aspects. Piacentio and Cardona [203] propose a robust optimization method for a CCHP system based on energetic analyses. They point out and verify that, by considering the energetic behavior, instead of improving the efficiency of an optimization algorithm, the optimization result can be significantly improved. Moran *et al.* [133] propose a thermoeconomic modeling approach, including the monthly operation cost, monthly fuel consumption, overall system efficiency, and so on, for micro-scale CCHP systems in residential use. This model helps to choose the optimal prime mover type and capacity by taking the ratio of required heating and cooling loads to the required electric loads into consideration.

In recent years, a matrix modeling approach based on the *energy hub* has begun to be used to model and optimize CCHP systems. In [36], the authors use a comprehensive approach to model a CCHP system in a compact matrix form. By adopting sequential quadratic programming (SQP), the power flow and electric cooling to cool load ratio are optimized. The case study in [36] shows that, compared with conventional operation strategies, the proposed optimal power flow and operation strategy can well control the CCHP system to achieve less PEC, ATC, and CDE. In [204], Chicco and Mancarella propose a matrix modeling approach for a small-scale trigeneration system and optimize the operation strategy for it. The model is built by introducing the concepts of efficiency matrices, dispatch factors, interconnection matrices, and input-to-output connectivity matrix. A depth-first manner is adopted to construct the overall plant efficiency matrix. In the optimization problem, NP techniques are used to obtain an optimal solution. In 2005, Geidl and Andersson [205] proposed a general matrix modeling and optimization approach for an energy system with various energy carriers. The optimization problem is a non-linear, multi-variable and inequality-constrained problem, which can be solved by the NP algorithm. In a later work [206], they use the concept of *energy hub* [207] to model the system by introducing the dispatch factors and coupling matrix. The dispatch and power flow are optimized by using the Karush–Kuhn–Tucker (KKT) conditions in order to minimize the total energy cost. The marginal cost is used to solve the KKT conditions. Using the same modeling approach, the optimization problem is solved by MATLAB *fmincon.m* in [208]. Ghaebi *et al.* [209] model a CCHP system using the TRR model to exergoeconomically optimize the cost of the total system production. Diverse parameters, including the air compressor pressure ratio, gas turbine inlet temperature, temperature in the heat recovery system, steam pressure, and so on, are involved in this model. The GA is adopted to solve for the optimal solution. Effects of the decision variables on different objective functions are also discussed in their work.

## 1.5.4  Sizing

Besides system configuration, operation strategy design, another equally important problem involved in the CCHP system is the facility sizing. As we all know, in the configuration step, one chooses prime movers in a vague way, that is, no accurate rated power is chosen. For instance, when designing a small-scale CCHP system whose capacity is in the range of 20 kW–1 MW, what is the specific value on earth? An appropriate operation strategy must depend on the facility size. Except for the PGU, other facilities' sizes can be determined by the required output. However, the PGU size should not be too small, which will make a CCHP system degrade to an SP system and will cause more electricity to be purchased from the local grid; this size should also not be too large, for high capital cost and low partial load efficiency. In many cases, the facility sizing problem can be inherently included in the modeling and optimization procedure. However, once the system configuration and operation strategy are designed, only the facility is to be sized to make the whole system efficient, economical and environmentally friendly. In [210], the authors examine the influence of different prime mover sizes and different operation strategies on the performance of the CCHP system. They validate three different sizes of the natural gas reciprocating engines under three different classical operation strategies, that is, FTL, FEL and following constant load (FCL). Different from other work, they use the actual market prices of electricity and natural gas instead of the flat one to minimize the CCHP performance indicators, including cost, PEC, and CDE, by optimizing the engine size. They point out that the optimal prime mover size would vary according to different evaluation criteria (EC). Sclafani and Beyene [211] discuss the challenges of matching and sizing for the CCHP system, due to the strong and frequently varying load conditions, from both the operation-related and weather-related aspects. In order to mitigate the part-load issues and address the system flexibilities, the load profiling strategies are adopted to optimize the selection and sizing of the prime mover. Liu *et al.* [181] adopt the enumeration algorithm to obtain the optimal size of the PGU under the proposed "balance"-space-based optimal operation strategy. In [164], the authors investigate the impact of the carbon tax on the sizing and operation strategy design of a medium-scale CHP system based on the IC engine. The thermoeconomic approach (annual cost flow approach) is used to optimize the IC engine capacity. In this work, under three operation modes, that is, one-way connection, two-way connection and heat demand following, the gas engine and diesel engine are sized to minimize the net annual cost. Shaneb *et al.* [212] model a $\mu$CHP plant with a generic deterministic LP model to minimize the annual cost. During the process of the optimization, the optimal size of the CHP unit and the size of the back-up heater can be obtained simultaneously. The sizing of the $\mu$CHP system can be completed either by the maximum rectangle method, which is to cover the average energy demand instead of the peak demand, or by LP. Different sizing results according to different prime movers, including IC engine, stirling engine, SOFC, and PEMFC, are also presented. Harrod *et al.* [213] provide a sizing analysis on the wood waste biomass-fired stirling-engine-based CCHP system in a small office building. From their analyses, the characteristics of the prime mover, including the capacity and electrical efficiency, have a significant influence on the cost and PEC of the CCHP system. They also point out that the optimal engine size, which aims to reduce the cost, is always larger than that of the reference system, which can result in a larger PEC. Thus, the trade-off between the cost and the PEC should be taken into consideration when determining the optimal engine size. In [214], Ren *et al.* adopt the MINLP to model a residential CHP system, which includes a storage tank and a back-up boiler. In the optimization process, the CHP system capacity is selected and the operating schedules are determined in order to minimize the annual overall cost of the energy system. They also analyze the sensitivity of the natural gas prices, electricity prices, carbon tax rates and electricity buy-back prices on the optimal CHP system capacity. Besides the prime mover, the capacity of the storage tank is also optimized to reach a good trade-off between the flexible storage management and heat loss to the surroundings. Zhang and Long [215] also formulate the sizing problem of the CHP system as an MINLP problem constrained by the energy demands, facility performance characteristics and the power flow in the whole system. The objective function is set to be the ATC by considering the operation strategy. In [216], a gas-fired grid-connected cogeneration system, which includes an absorption chiller and a thermal storage system, is modeled as an MINLP model. The Newton–Raphson and conjugate method with tangent estimates and forward derivatives are used to search for the optimal facility size aiming to minimize the operational cost. The best feasible practical solution is determined by the dynamic programming principle. Financial analysis, including the payback period and internal rate of return, is also reported in this work. Similar to [190], Rubio-Maya *et al.* [217] use the superstructure concept to choose the optimal size, which includes the net power of the prime mover, cooling capacity of the thermally activated facility and fresh water capacity of the desalting unit, of a polygeneration system. The objective function is set to be the net present value, which is equally constrained by the energy and mass balance, and unequally constrained by the PES, GHG emissions and legal framework aspects, and is solved as an MINLP problem. Sensitive analyses based on the prices of electricity, fuel and water are also provided in this work. In [218], the aggregate thermal demand method (ATD$_e$ method) is adopted to estimate the PES. In addition, the estimation of PES, which is used for the sizing problem, is based on the annual value. The PES strategy can combine both the economic and environmental considerations. The most advantageous aspect of this work is the simple calculation method, in which only a few representative global data are needed. In [219], using the concept of energy hub, Sheikhi *et al.* model the CCHP system optimization problem by an NP model. Aiming to achieve the maximum net benefit, the operation point of the energy hub and the size of prime mover, absorption chiller, auxiliary boiler and heating storage devices are optimized by solving the non-linear programming in GAMS software. In addition, a financial analysis, which includes the net present value and internal rate of return, is conducted using the technique of discounted cash flow analysis.

## 1.6    Development and Barriers of CHP/CCHP Systems in Representative Countries

### 1.6.1    The United States

The US government began to develop CHP/CCHP plants since 1978, when the PURPA was proposed. In the PURPA, utilities are required to interconnect with and purchase electricity from cogeneration systems, in order to give industrial and institutional users access to the grid and allow excess electricity to be sold back. With the help of the PURPA and the federal tax credit for CHP investment, the installed capacity of CHP/CCHP systems grew to 45 GW in 1995 from 12 GW in 1980. Due to the intense competition and instability in the electricity market, the development of CHP/CCHP plants slowed down in the 1990s. There was only 1 GW installed capacity increase from 1995 to 1998. To boost the development, together with the Environmental Protection Agency (EPA), the US Department of Energy (DOE) proposed the "Combined cooling heating & power for buildings 2020 vision", which aimed to double the installed capacity in 2010. Following the proposed document, the installed capacity grew significantly to 56 GW in 2001. Then in 2004, with a total installed capacity of 80 GW, the goal of 92 GW had been almost achieved. In 2009, after the Energy Policy Act in 2005, the installed capacity had reached 91 GW. The development trend of the US CHP/CCHP installed capacity from 1970 is shown in Figure 1.6 [220]. As of 2011, as shown in Figure 1.7, 30% CHP/CCHP installed capacity is used in chemical industries, 17% is used for petroleum refining, 14% for paper industries, 12% for commercial or institutional buildings, 8% for food manufacturing, 8% for other manufacturing, 5% for primary metals industries, and 6% for other industries. According to "The White Paper on CHP in a Clean Energy Standard" [221], the US DOE aims to have an 11% increase, from the current 9%, of

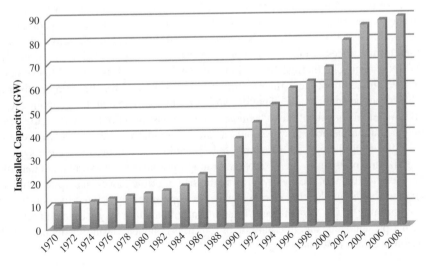

**Figure 1.6**    US CHP/CCHP development from 1970 [220]

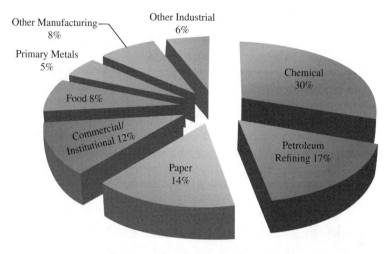

**Figure 1.7**    The installed capacity of CHP/CCHP plants classified by applications in the US

CHP share of the US electric power by 2030. By doing so, a 60% projected increase in US carbon emissions can be avoided; over 1 million new, highly skilled jobs can be created; and $234 billion in new investment can be generated.

However, there still exist some barriers to further development of CHP/CCHP plants in the US. The first one is the high capital investment of CHP/CCHP plants. A firm may not have sufficient budget to invest in such a high capital cost plant; or, if not sure about the payback of such a plant, it still cannot invest in it. Secondly, to keep a connection with the utility grid to supply power needs beyond the self generation capacity, extra charges will be made for this connection. This will no doubt reduce the money-saving potential of CHP/CCHP plants. Thirdly, non-uniform interconnection standards make it difficult for manufacturers to provide CHP/CCHP components. In addition, some policies, such as the Clean Air Act's New Source Review, only consider short-term carbon emissions instead of a long-term and overall vision. Because "the CHP/CCHP can increase onsite air emissions even as it reduces total emissions associated with the facilities heat and electricity consumption" [222], the development of CHP plants can be restricted by such regulations. Finally, further theoretical research, development, and demonstrations should be conducted to find out more efficient operation modes, system configurations, and system components.

## 1.6.2    The United Kingdom

In the UK, the number and installed capacity of CHP/CCHP plants increased dramatically from 1999 to 2000, during which time the UK government used methods of fiscal incentives, grant support, regulatory framework, promotion of innovation, and government leadership and partnership to support the development of CHP/CCHP. Before 2000, the installed capacity kept around 3.5 GW, while in 2000, it increased

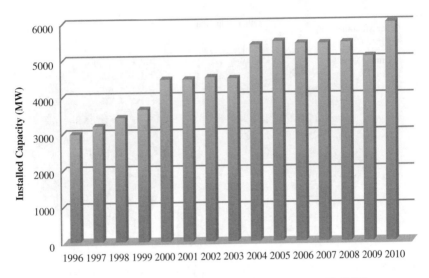

**Figure 1.8**   The CHP/CCHP installed capacity in the UK [223]

to 4.5 GW. From then on, the UK government continually drafted a series of policies to target achieving 10 GW *of good quality* installed CHP plants. By the end of 2010, the total installed capacity in the UK reached 6 GW, as shown in Figure 1.8 [223]. Classified by applications, as shown in Figure 1.9 [223], 38% of the installed capacity is used for oil and gas terminals and refineries, another 30% is used for

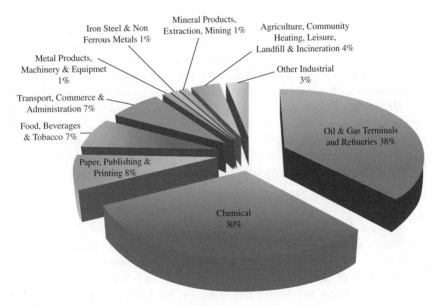

**Figure 1.9**   The installed capacity of CHP plants classified by applications in the UK [223]

chemical industries and only 4% is used for community usage, and so on. In [224], it is also pointed out that "the *of good quality* CHP will be a key technology in helping to deliver our carbon budgets while the grid decarbonizes, and will still play a pivotal role in providing secure and cost-effective energy supplies, particularly for industry. The government will continue to promote the development of *of good quality* CHP in the UK".

Meanwhile, there are still some obstacles for CHP/CCHP to be further developed in the UK. The first one is the inconsistency between incentive frameworks and market signals. One significant characteristic of the UK market is price volatility. The differential between electricity and gas prices put the investment of CHP/CCHP in an uncertain situation. This issue may be addressed by the Climate Change Levy. Secondly, in theory, the establishment of the carbon market should directly support the expansion of CHP/CCHP capacity. However, due to the unstable carbon price and uncertain allocation arrangements of CHP/CCHP plants, this theory has not yet been verified. Only with a robust price signal and a stable carbon market, the direct relationship between the CHP/CCHP expansion and carbon market can be established. In addition, lacking locational signals for heat utilization also restricts the development of CHP/CCHP plants. Moreover, to achieve the peak efficiency, the heat transmission and distribution network should be completed. Finally, insufficient incentive to invest in heat distribution infrastructures also slows down the pace of CHP/CCHP development in the UK.

## 1.6.3 The People's Republic of China

Due to the Reform and Open Up to the Outside World Policy, the rapid development of economics, technology, and industry leads China to be the well known second-largest energy consumer and carbon emitter in the world. To solve the problem of the increasing demand for primary energy, China has issued a series of policies, including the Energy Saving Law, the Renewable Energy Law, the Air Pollution Prevention Law, and the Environment Protection Law, to support the development of CHP/CCHP plants since the 1980s. In addition, accompanying those laws, some standards, for example, Energy Efficiency Standards for Buildings, and Energy Efficiency Standards for Appliances, and some dedicated funds, subsidies and discounted loans for energy efficiency investments have been implemented by the Chinese government. These steps make China the second-largest country in terms of installed CHP capacity. In 1986, the Notice on the Report Regarding the Work on Strengthening Urban District Heat Supply Management enhanced the urban district heating supply management. The China Energy Conservation Law, drafted in 1997, listed CHP as a key national energy conservation technology that should be encouraged. The 1998 Some Regulations for CHP Development considered the ratio between heat and electricity as an important indicator to define and approve new CHP. In 2004, the China Medium- and Long-Term Energy Development Plan considered CHP/DHC as an encouraging technology and named CHP as one of the 10 key national energy conservation programs. In 2006, the NDRC's China Energy Conservation Technology Policy Outline recommended that CHP should take the place of small heating boilers; and they should be developed in large- and medium-sized cities in north heating areas. The 2007 Implementation Scheme of the National 10 Key Energy Conservation Projects further specified important applications and supporting policies for CHP. Another important policy that could boost the development of CHP/CCHP plants in China is the Industrial Guidance Catalogue for Foreign Investments drafted in 2007. This policy encouraged foreign investments and operations of CHP/CCHP power stations in China.

In 1990, the total installed CHP capacity in China was only 10 GW. After 10 years' construction and development, with an annual growth rate of 11.6%, a goal of 30 GW installed capacity was achieved in 2000. By 2005, almost 70 GW of capacity had been installed. Up to 2006, over 2600 CHP plants with over 80 GW capacity had been installed in China. The development trend is shown in Figure 1.10 [225]. The share of CHP capacity in thermal power generation in China is shown in Figure 1.11 [225].

With no doubt, CHP/CCHP is a promising solution for the energy shortages and air pollution problem, however, there are still some barriers to further developing CHP/CCHP plants in China. The first urgent problem to be solved is to reform the energy price policies. In China, even though the coal price, which has increased dramatically, is based on the market, the price of electricity, which has slightly increased, is decided by the government. Unbalanced increasing rates between the prices of coal and electricity severely restrict the development of CHP/CCHP in China. Besides the reform of energy price policies, heating and power sector reforms also need to be taken into consideration. Not only the economic and price aspect, but also some favorable fiscal and tax incentives should be proposed to support the construction of CHP/CCHP. In addition, since some newly built CHP projects are operating only in

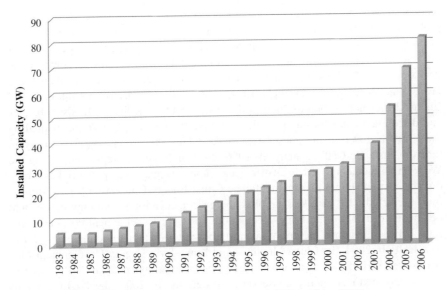

**Figure 1.10**     The installed capacity of CHP in China [225]

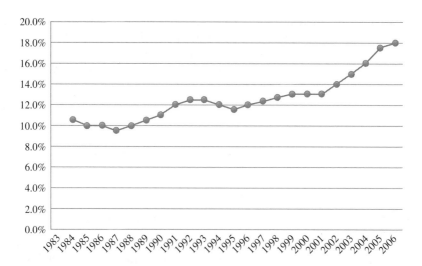

**Figure 1.11**     Share of CHP capacity in thermal power generation [225]

thermal generation mode after being established, energy efficiencies of these plants are significantly reduced. Thus, the monitoring and enforcement of the government energy price policies should be enhanced. Finally, due to an increasing number of large and more efficient CHP plants, some old, small but quite efficient CHP plants are being forced to shut down. Hence, policies that are suitable for the small but efficient units should be drafted to keep the overall efficiency.

## 1.7   Summary

This chapter elaborately presents the state-of-the-art of CCHP systems. The CCHP, which can provide the cooling energy by adopting thermally activated technology, is a concept extended from the CHP system. To construct an economical and efficient CCHP system, the type of facilities should be firstly determined according to local resources, and current and future energy market. Different types of prime movers and thermally activated technologies are introduced. The system configuration varies according to different usages, including commercial buildings, residential buildings, supermarkets, universities, hospitals, and so on. Some classical CHP/CCHP configurations classified by prime mover types and system capacities are also introduced in this chapter. With the determined facility type and system configuration, to achieve the optimal operation status, the operation strategy, power flow, and the facilities capacity should be optimized. The current research on the optimal operation strategy design, power flow optimization, and sizing problem are reviewed in this chapter. As many countries have begun to develop CHP/CCHP systems, the development history and current status of three representative countries, that is, the US, the UK, and the People's Republic of China, are presented. From these three countries, where the situations are different, some similar obstacles and solutions are discussed to further develop CCHP technology for the whole world. The high capital cost, government supporting policies, favorable fiscal and tax incentives, and further research and development should receive more attention. In the future development of CHP/CCHP systems, incorporating renewable energy is urged.

# References

[1] I. Tatsumi, "Applicability of micro gas turbine as distributed generation and its connection to utility power grid," *Journal of Gas Turbine Society of Japan*, vol. 29, no. 3, pp. 141–145, 2001.

[2] A. C. Oliveira, C. Afonso, J. Matos, S. Riffat, M. Nguyen, and P. Doherty, "A combined heat and power system for buildings driven by solar energy and gas," *Applied Thermal Engineering*, vol. 22, no. 6, pp. 587–593, 2002.

[3] N. Lior and N. Zhang, "Energy, exergy, and second law performance criteria," *Energy*, vol. 32, no. 4, pp. 281–296, 2007.

[4] T. G. Bourgeoisa, B. Hedmanb, and F. Zalcmanc, "Creating markets for combined heat and power and clean distributed generation in New York State," *Environmental Pollution*, vol. 123, no. 3, pp. 451–462, 2003.

[5] Y. Huangfu, J. Wu, R. Wang, X. Kong, and B. Wei, "Evaluation and analysis of novel micro-scale combined cooling, heating and power (MCCHP) system," *Energy Conversion and Management*, vol. 48, no. 5, pp. 1703–1709, 2007.

[6] E. Bilgen, "Exergetic and engineering analyses of gas turbine based cogeneration systems," *Energy*, vol. 25, no. 12, pp. 1215–1229, 2000.

[7] X. Q. Kong, R. Z. Wang, and X. H. Huang, "Energy optimization model for a CCHP system with available gas turbine," *Applied Thermal Engineering*, vol. 25, no. 2–3, pp. 377–391, 2005.

[8] K. Khana, M. Rasulb, and M. Khan, "Energy conservation in buildings: cogeneration and cogeneration coupled with thermal energy storage," *Applied Energy*, vol. 77, no. 1, pp. 15–34, 2004.

[9] V. Havelsky, "Energetic efficiency of cogeneration systems for combined heat, cold and power production," *International Journal of Refrigeration*, vol. 22, no. 6, pp. 479–485, 1999.

[10] D. W. Wu and R. Z. Wang, "Combined cooling, heating and power: A review," *Progress in Energy and Combustion Science*, vol. 32, no. 5–6, pp. 459–495, 2006.

[11] J. Xu, J. Sui, B. Li, and M. Yang, "Research, development and the prospect of combined cooling, heating, and power systems," *Energy*, vol. 35, no. 11, pp. 4361–4367, 2010.

[12] J. Hernandez-Santoyoa and A. Sanchez-Cifuentes, "Trigeneration: an alternative for energy savings," *Applied Energy*, vol. 76, no. 1–3, pp. 119–127, 2003.

[13] Energetics Incorporated, "Market assessment of distributed energy in new commercial and institutional buildings and critical infrastructure facilities," Energetics Incorporated, Tech. Rep., 2006.

[14] L. Dong, H. Liu, and S. Riffat, "Development of small-scale and micro-scale biomass-fuelled CHP systems—a literature review," *Applied Thermal Engineering*, vol. 29, no. 11–12, pp. 2119–2126, 2009.

[15] M. D. d'Accadia, M. Sasso, S. Sibilio, and L. Vanoli, "Micro-combined heat and power in residential and light commercial applications," *Applied Thermal Engineering*, vol. 23, no. 10, pp. 1247–1259, 2003.

[16] R. H. Lasseter and P. Paigi, "Microgrid: A conceptual solution," in *Proceedings of IEEE Power Electronics Specialist Conference*, vol. 6, Aachen, Germany, June 20–25, 2004, pp. 4285–4290.

[17] A. Martens, "The energetic feasibility of CHP compared to the separate production of heat and power," *Applied Thermal Engineering*, vol. 18, no. 11, pp. 935–946, 1998.

[18] O. Balli, H. Aras, and A. Hepbasli, "Exergoeconomic analysis of a combined heat and power (CHP) system," *International Journal of Energy Research*, vol. 32, no. 4, pp. 273–289, 2008.

[19] P. A. Pilavachi, C. P. Roumpeas, S. Minett, and N. H. Afgan, "Multi-criteria evaluation for CHP system options," *Energy Conversion and Management*, vol. 47, no. 20, pp. 3519–3529, 2006.

[20] N. Fumo, P. J. Mago, and L. M. Chamra, "Cooling, heating, and power energy performance for system feasibility," in *Proceedings of the Institution of Mechanical Engineers, Part A: Journal of Power and Energy*, vol. 222, no. 4, 2008, pp. 347–354.

[21] J. M. Carrasco, L. G. Franquelo, J. T. Bialasiewicz, E. Galvan, R. C. P. Guisado, M. A. M. Prats, J. I. Leno, and N. Moreno-Alfonso, "Power-electronic systems for the grid integration of renewable energy sources: A survey," *IEEE Transactions on Industrial Electronics*, vol. 53, no. 4, pp. 1002–1016, 2006.

[22] D. Weisser and R. S. Garcia, "Instantaneous wind energy penetration in isolated electricity grids: Concepts and review," *Renewable Energy*, vol. 30, no. 8, pp. 1299–1308, 2005.

[23] I. Dincer, "Renewable energy and sustainable development: a crucial review," *Renewable & Sustainable Energy Reviews*, vol. 4, no. 2, pp. 157–175, 2000.

[24] US Department of Energy. (2011) Types of fuel cells. [Online]. Available: http://energy.gov/eere/fuelcells/types-fuel-cells. Accessed November 10, 2016.

[25] P. Agostini, M. Botteon, and C. Carraro, "A carbon tax to reduce $CO_2$ emissions in Europe," *Energy Economics*, vol. 14, no. 4, pp. 279–290, 1992.

[26] N. Floros and A. Viachou, "Energy demand and energy-related $CO_2$ emissions in Greek manufacturing: Assessing the impact of a carbon tax," *Energy Economics*, vol. 27, no. 3, pp. 387–413, 2005.

[27] S. Cnossen, "Tax policy in the European Union: a review of issues and options," *FinanzArchiv/Public Finance Analysis*, vol. 58, no. 4, pp. 466–558, 2001.

[28] A. Bernard, M. Vielle, and L. Viguier, "Carbon tax and international emissions trading: A Swiss perspective," *Advances in Global Change Research*, vol. 22, pp. 295–319, 2005.

[29] R. Shrestha, S. Pradhan, and M. H. Liyanage, "Effects of carbon tax on greenhouse gas mitigation in Thailand," *Climate Policy*, vol. 8, no. 1, pp. S140–S155–, 2008.

[30] K. Alanne and A. Saari, "Distributed energy generation and sustainable development," *Renewable & Sustainable Energy Reviews*, vol. 10, no. 6, pp. 539–558, 2006.

[31] J. Driesen, F. Katiraei, and K. U. Leuven, "Design for distributed energy resources," *IEEE Power & Energy Magazine*, vol. 6, no. 3, pp. 30–40, 2008.

[32] J. Wang, Z. Zhai, Y. Jing, and C. Zhang, "Particle swarm optimization for redundant building cooling heating and power system," *Applied Energy*, vol. 87, no. 12, pp. 3668–3679, 2010.

[33] J. Wang, Y. Jing, and C. Zhang, "Optimization of capacity and operation for CCHP system by genetic algorithm," *Applied Energy*, vol. 87, no. 4, pp. 1325–1335, 2010.

[34] L. Yong and R. Z. Wang, "Adsorption refrigeration: A survey of novel technologies," *Recent Patents on Engineering*, vol. 1, no. 1, pp. 1–21, 2007.

[35] J. A. Jones, "Sorption refrigeration research at JPL/NASA," *Heat Recovery Systems and CHP*, vol. 13, no. 4, pp. 363–371, 1993.

[36] M. Liu, Y. Shi, and F. Fang, "Optimal power flow and PGU capacity of CCHP systems using a matrix modelling approach," *Applied Energy*, vol. 102, pp. 794–802, 2013.

[37] M. Goodell. (2011) The advantages of cogeneration and trigeneration. [Online]. Available: http://www.trigeneration.com/. Accessed November 10, 2016.

[38] P. Arcuri, G. Florio, and P. Fragiacomo, "A mixed integer programming model for optimal design of trigeneration in a hospital complex," *Energy*, vol. 32, no. 8, pp. 1430–1447, 2010.

[39] Y. Ge, S. Tassou, I. Chaer, and N. Suguartha, "Performance evaluation of a trigeneration system with simulation and experiment," *Applied Energy*, vol. 86, no. 11, pp. 2317–2326, 2009.

[40] G. G. Maidment and R. M. Tozer, "Combined cooling heat and power in supermarkets," *Applied Thermal Engineering*, vol. 22, no. 6, pp. 653–665, 2002.

[41] L. Gao, H. Wu, H. Jin, and M. Yang, "System study of combined cooling, heating and power system for eco-industrial parks," *International Journal of Energy Research*, vol. 32, no. 12, pp. 1107–1118, 2008.

[42] WADE, "4,600 kWe gas turbine CCHP Shanghai Pudong International Airport," WADE, Tech. Rep., 2011.

[43] D. Huang, "Assessment on barriers of CHP/trigeneration promotion and potential countermeasure in China," Zhejiang Energy Research Institute (ZERI), Tech. Rep., 2004.

[44] H. Li, L. Fu, K. Geng, and Y. Jiang, "Energy utilization evaluation of CCHP systems," *Energy and Buildings*, vol. 38, no. 3, pp. 253–257, 2006.

[45] Wikipedia. (2011) Reciprocating engine. [Online]. Available: http://en.wikipedia.org/wiki/Reciprocating_engine. Accessed February 5, 2017.

[46] Energy and Environmental Analysis, Inc., *"Technology characterization: reciprocating engines,"* Energy and Environmental Analysis, Inc., Tech. Rep., 2008.

[47] H. Onovwiona and V. Ugursal, "Residential cogeneration systems: Review of the current technology," *Renewable & Sustainable Energy Reviews*, vol. 10, no. 5, pp. 389–431, 2005.

[48] I. Knight and V. Ugursal, "Residential cogeneration systems: A review of the current technologies, a report of annex 42 of the international energy agency, energy conservation in buildings and community systems programme," Tech. Rep., 2005.

[49] H. I. Onovwiona, V. I. Ugursal, and A. S. Fung, "Modeling of internal combustion engine based cogeneration systems for residential applications," *Applied Thermal Engineering*, vol. 27, no. 5–6, pp. 848–861, 2007.

[50] HONDA Motor Co. (2001) Honda to begin practical tests of small household cogeneration unit. [Online]. Available: http://world.honda.com/news/2001/p010830-eng.html. Accessed February 5, 2017.

[51] G. Bidini, U. Desideri, S. Saetta, and P. P. Bocchini, "Internal combustion engine combined heat and power plants: Case study of the University of Perugia power plant," *Applied Thermal Engineering*, vol. 18, no. 6, pp. 401–412, 1998.

[52] A. A. Jalalzadeh-Azar, S. Slayzak, R. Judkoff, T. Schamuser, and R. DeBlasio, "Performance assessment of a desiccant cooling system in a CHP application incorporating an IC engine," *International Journal of Distributed Energy Resources*, vol. 1, no. 2, pp. 163–184, 2004.

[53] G. A. Longo, A. Gasparella, and C. Zilio, "Analysis of an absorption machine driven by the heat recovery on an I.C. reciprocating engine," *International Journal of Energy Research*, vol. 29, no. 8, p. 2005–, 2005.

[54] M. Talbi and B. Agnew, "Energy recovery from diesel engine exhaust gases for performance enhancement and air conditioning," *Applied Thermal Engineering*, vol. 22, no. 6, pp. 693–702, 2002.

[55] J. M. Riley and S. D. Probert, "Carbon-dioxide emissions from an integrated small-scale CHP and absorption chiller system," *Applied Energy*, vol. 61, no. 4, pp. 193–207, 1998.

[56] Wikipedia. (2011) Internal combustion engine. [Online]. Available: http://en.wikipedia.org/wiki/Internal_combustion_engine. Accessed February 5, 2017.

[57] Energy and Environmental Analysis, Inc., "Technology characterization: gas turbines," Energy and Environmental Analysis, Inc., Tech. Rep., 2008.

[58] A. Poullikkas, "An overview of current and future sustainable gas turbine technologies," *Renewable & Sustainable Energy Reviews*, vol. 9, no. 5, pp. 409–443, 2005.

[59] P. Pilavachi, "Power generation with gas turbine systems and combined heat and power," *Applied Thermal Engineering*, vol. 20, no. 15–16, pp. 1421–1429, 2000.

[60] G. Maidment, X. Zhao, and S. Riat, "Combined cooling and heating using a gas engine in a supermarket," *Applied Energy*, vol. 68, no. 4, pp. 321–335, 2001.

[61] D.-C. Sue and C.-C. Chuang, "Engineering design and exergy analyses for combustion gas turbine based power generation system," *Energy*, vol. 29, no. 8, pp. 1183–1205, 2004.

[62] X. Q. Kong, R. Z. Wang, J. Y. Wu, X. H. Huang, Y. Huangfu, D. W. Wu, and Y. Xu, "Experimental investigation of a micro-combined cooling, heating and power system driven by a gas engine," *International Journal of Refrigeration*, vol. 28, no. 7, pp. 977–987, 2005.

[63] Wikipedia. (2011) Steam turbine. [Online]. Available: http://en.wikipedia.org/wiki/Steam_turbine. Accessed February 5, 2017.

[64] Energy and Environmental Analysis, Inc., "Technology characterization: steam turbines," Energy and Environmental Analysis, Inc., Tech. Rep., 2008.

[65] Energy and Environmental Analysis, Inc., "Technology characterization: microturbines," Energy and Environmental Analysis, Inc., Tech. Rep., 2008.

[66] Energy Nexus Group, "Technology characterization: Microturbines," Energy Nexus Group, Tech. Rep., 2002.

[67] P. A. Pilavachi, "Mini- and micro-gas turbines for combined heat and power," *Applied Thermal Engineering*, vol. 22, no. 18, pp. 2003–2014, 2002.

[68] O. Balli and H. Aras, "Energetic and exergetic performance evaluation of a combined heat and power system with the micro gas turbine (MGTCHP)," *International Journal of Energy Research*, vol. 31, no. 14, pp. 1425–1440, 2007.

[69] S. A. Tassou, I. Chaer, N. Sugiartha, Y. T. Ge, and D. Marriott, "Application of tri-generation systems to the food retail industry," *Energy Conversion and Management*, vol. 48, no. 11, pp. 2988–2995, 2007.

[70] S. Karellas, J. Karl, and E. Kararas, "An innovative biomass gasification process and its coupling with microturbine and fuel cell systems," *Energy*, vol. 33, no. 2, pp. 284–291, 2008.

[71] D. T. Rizy, A. Zaltash, S. D. Labinov, A. Y. Petrov, and P. D. Fairchild, "DER performance testing of a microturbine-based combined cooling, heating, and power (CHP) system," in *Proceedings of Power System 2002 Conference "Impact of Distributed Generation"*, Clemson, SC, USA, March 13–15, 2002.

[72] J. C. Bruno, V. Ortega-López, and A. Coronas, "Integration of absorption cooling systems into micro gas turbine trigeneration systems using biogas: case study of a sewage treatment plant," *Applied Energy*, vol. 86, no. 6, pp. 837–847, 2009.

[73] Y. Hwang, "Potential energy benefits of integrated refrigeration system with microturbine and absorption chiller," *International Journal of Refrigeration*, vol. 27, no. 8, pp. 816–829, 2004.

[74] S. Velumani, C. E. Guzman, R. Peniche, and R. Vega, "Proposal of a hybrid CHP system: SOFC/microturbine/absorption chiller," *International Journal of Energy Research*, vol. 34, no. 12, pp. 1088–1095, 2010.

[75] X. Liao and R. Rademacher, "Absorption chiller crystallization control strategies for integrated cooling heating and power systems," *International Journal of Refrigeration*, vol. 30, no. 5, pp. 904–911, 2007.

[76] N. Sugiartha, S. A. Tassou, I. Chaer, and D. Marriott, "Trigeneration in food retail: An energetic, economic and environmental evaluation for a supermarket application," *Applied Thermal Engineering*, vol. 29, no. 13, pp. 2624–2632, 2009.

[77] M. Medrano, J. Mauzey, V. McDonell, S. Samuelsen, and D. Boer, "Theoretical analysis of a novel integrated energy system formed by a microturbine and an exhaust fired single-double effect absorption chiller," *International Journal of Thermodynamics*, vol. 9, no. 1, pp. 29–36, 2006.

[78] A. Vidal, J. C. Bruno, R. Best, and A. Coronas, "Performance characteristics and modelling of a micro gas turbine for their integration with thermally activated cooling technologies," *International Journal of Energy Research*, vol. 31, no. 2, pp. 119–134, 2007.

[79] A. Huicochea, W. Rivera, G. Gutiérrez-Urueta, J. C. Bruno, and A. Coronas, "Thermodynamic analysis of a trigeneration system consisting of a micro gas turbine and a double effect absorption chiller," *Applied Thermal Engineering*, vol. 31, no. 16, pp. 3347–3353, 2011.

[80] I. Obernberger, H. Carlsen, and F. Biedermann, "State-of-the-art and future developments regarding small-scale biomass CHP systems with a special focus on ORC and stirling engine technologies," in *Proceedings of International Nordic Bioenergy Conference*, 2003, pp. 1–7.

[81] Wikipedia. (2012) Stirling engine. [Online]. Available: https://en.wikipedia.org/wiki/Stirling_engine. Accessed February 5, 2017.

[82] J. Harrison, "Micro combined heat & power," EA Technology, Tech. Rep., 2002.

[83] C. S. Vineeth, *Stirling Engines: A Beginners Guide*. C. S. Vineeth, 2011.

[84] R. Z. W. X. Q. Kong and X. H. Huang, "Energy efficiency and economic feasibility of CCHP driven by stirling engine," *Energy Conversion and Management*, vol. 45, no. 9–10, pp. 1433–1442, 2004.

[85] F. Biedermann, H. Carlsen, M. Schoch, and I. Obernberger, "Operating experiences with a small-scale CHP pilot plant based on a 35 kW$_{EL}$ hermetic four cyclinder stirling engine for biomass fuels," BIOS BIOENERGIESYSTEME GmbH, Tech. Rep., 2004.

[86] A. A. Aliabadi, M. J. Thomson, J. S. Wallace, T. Tzanetakis, W. Lamont, and J. D. Carlo, "Efficiency and emissions measurement of a stirling-engine-based residential microcogeneration system run on diesel and biodiesel," *Energy & Fuels*, vol. 23, no. 2, pp. 1032–1039, 2009.

[87] D. Scarpete, K. Uzuneanu, and N. Badea. "Stirling engine in residential systems based on renewable energy," in *Advances in Energy Planning*, Environmental Education and Renewable Energy

Resources, Tunisia, 2010, pp. 124–129. Available: http://www.wseas.us/e-library/conferences/ 2010/Tunisia/EPERES/EPERES-20.pdf. Accessed February 9, 2017.

[88] R. S. Khurmi and R. S. Sedha, *Material Science*. S. Chand & Company Ltd, 2010.

[89] C. Wang and M. H. Nehrir, "Distributed generation applications of fuel cells," in *Proceedings of Power Systems Conference: Advanced Metering, Protection, Control, Communication, and Distributed Resources*, Clemson, SC, USA, March 14–17, 2006, pp. 244–248.

[90] A. B. Stambouli and E. Traversa, "Fuel cells, an alternative to standard sources of energy," *Renewable & Sustainable Energy Reviews*, vol. 6, no. 3, pp. 295–304, 2002.

[91] K. Kordesch and G. Simader, *Fuel Cells and Their Applications*. Wiley-VCH, 1996.

[92] L. K. C. Tse, S. Wilkins, N. McGlashan, B. Urban, and R. Martinez-Botas, "Solid oxide fuel cell/gas turbine trigeneration system for marine applications," *Journal of Power Sources*, vol. 196, no. 6, pp. 3149–3162, 2011.

[93] P. Kazempoor, V. Dorer, and F. Ommi, "Modelling and performance evaluation of solid oxide fuel cell for building integrated co- and polygeneration," *Fuel Cells*, vol. 10, no. 6, pp. 1074–1094, 2010.

[94] E. Baniasadi and A. A. Alemrajabi, "Fuel cell energy generation and recovery cycle analysis for residential application," *International Journal of Hydrogen Energy*, vol. 35, no. 17, pp. 9460–9467, 2010.

[95] V. Verda and M. C. Quaglia, "Solid oxide fuel cell systems for distributed power generation and cogeneration," *International Journal of Hydrogen Energy*, vol. 33, no. 8, pp. 2087–2096, 2008.

[96] F. A. Al-Sulaiman, I. Dincer, and F. Hamdullahpur, "Exergy analysis of an integrated solid oxide fuel cell and organic Rankine cycle for cooling, heating and power production," *Journal of Power Sources*, vol. 195, no. 8, pp. 2346–2354, 2010.

[97] F. A. Al-Sulaiman, I. Dincer, and F. Hamdullahpur, "Energy analysis of a trigeneration plant based on solid oxide fuel cell and organic Rankine cycle," *International Journal of Hydrogen Energy*, vol. 35, no. 10, pp. 5104–5113, 2010.

[98] Z. Yu, J. Han, X. Cao, W. Chen, and B. Zhang, "Analysis of total energy system based on solid oxide fuel cell for combined cooling and power applications," *International Journal of Hydrogen Energy*, vol. 35, no. 7, pp. 2703–2707, 2010.

[99] P. Kazempoor, V. Dorer, and F. Ommi, "Evaluation of hydrogen and methane-fuelled solid oxide fuel cell systems for residential applications: system design alternative and parameter study," *International Journal of Hydrogen Energy*, vol. 34, no. 20, pp. 8630–8644, 2009.

[100] I. Malico, A. P. Carvalhinho, and J. Tenreiro, "Design of a trigeneration system using a high-temperature fuel cell," *International Journal of Energy Research*, vol. 33, no. 2, pp. 144–151, 2009.

[101] R. J. Braun, S. A. Klein, and D. T. Reindl, "Evaluation of system configurations for solid oxide fuel cell-based micro-combined heat and power generators in residential applications," *Journal of Power Sources*, vol. 158, no. 2, pp. 1290–1305, 2006.

[102] C. Weber, M. Koyama, and S. Kraines, "$CO_2$-emissions reduction potential and costs of a decentralized energy system for providing electricity, cooling and heating in an office-building in Tokyo," *Energy*, vol. 31, no. 14, pp. 3041–3061, 2006.

[103] W. G. Colella and R. Timme, "Optimizing operation of stationary fuel cell systems (FCS) within district cooling and heating networks," in *ASME Conference Proceedings*, vol. 263, 2010, pp. 1–114.

[104] P. Margalef and S. Samuelsen, "Integration of a molten carbonate fuel cell with a direct exhaust absorption chiller," *Journal of Power Sources*, vol. 195, no. 17, pp. 5674–5685, 2010.

[105] G. Bizzarri, "On the size effect in PAFC grid-connected plant," *Applied Thermal Engineering*, vol. 26, no. 10, pp. 1001–1007, 2006.

[106] J. Deng, R. Wang, and G. Han, "A review of thermally activated cooling technologies for combined cooling, heating and power systems," *Progress in Energy and Combustion Science*, vol. 37, no. 2, pp. 172–203, 2011.

[107] W. B. Gosney, *Principle of refrigeration*. Cambridge University Press, 1982.

[108] P. Srikhirim, S. Aphornratana, and S. Chungpaibulpatana, "A review of absorption refrigeration technologies," *Renewable & Sustainable Energy Reviews*, vol. 5, no. 4, pp. 343–372, 2001.

[109] R. A. Marcriss, J. M. Gutraj, and T. S. Zawacki, "Absorption fluid data survey: final report on worldwide data," Institute of Gas Technology, Tech. Rep., 1988.

[110] I. Stambler, "4.6 MW plant with an indirect fired 2600 ton chiller at 76.8% efficiency," *Gas Turbine World*, vol. 34, no. 4, pp. 14–17, 2004.

[111] A. Marantan, "Optimization of intergrated micro-turbine and absorption chiller systems in CHP for buildings application," PhD dissertation, University of Maryland, 2002.

[112] J. Bassols, B. Kuchelkorn, J. Langreck, R. Schneder, and H. Veelken, "Trigeneration in the food industry," *Applied Thermal Engineering*, vol. 22, no. 6, pp. 595–602, 2002.

[113] R. E. Critoph and Y. Zhong, "Review of trends in solid sorption refrigeration and heat pumping technology," *Proceedings of the Institution of Mechanical Engineers, Part E: Journal of Process Mechanical Engineering*, vol. 219, no. 3, pp. 285–300, 2005.

[114] C. A. Balaras, G. Grossman, H. Henning, C. A. I. Ferreria, E. Podesser, L. Wang, and E. Wiemken, "Solar air conditioning in Europe—an overview," *Renewable & Sustainable Energy Reviews*, vol. 11, no. 2, pp. 299–314, 2007.

[115] D. S. Kim and C. A. I. Ferreria, "Solar refrigeration options—a state-of-the-art review," *International Journal of Refrigeration*, vol. 3, no. 1, pp. 3–15, 2008.

[116] R. Z. Wang, J. Y. Wu, Y. X. Xu, and W. Wang, "Performance researches and improvements on heat regenerative adsorption refrigerator and heat pump," *Energy Conversion and Management*, vol. 42, no. 2, pp. 233–249, 2001.

[117] S. Li and J. Y. Wu, "Theoretical research of a silica gel-water adsorption chiller in a micro combined cooling, heating and power (CCHP) system," *Applied Energy*, vol. 86, no. 6, pp. 958–967, 2009.

[118] Midwest CHP Application Center (University of Illinois at Chicago) and Avalon Consulting, Inc., *Combined Heat and Power Source Guide*. Midwest CHP Application Center (University of Illinois at Chicago) and Avalon Consulting, Inc., 2003.

[119] Mississippi Cooling, Heating, and Power (Micro-CHP) and Bio-fuel Center, "Cooling, heating, and power for buildings (CHP-B) instructional module," Department of Mechanical Engineering, Mississippi State University, Tech. Rep., 2004.

[120] H.-M. Henning, T. Pagano, S. Mola, and E. Wiemken, "Micro tri-generation system for indoor air conditioning in the mediterranean climate," *Applied Thermal Engineering*, vol. 27, no. 13, pp. 2188–2194, 2007.

[121] L. Fu, X. L. Zhao, S. G. Zhang, Y. Jiang, H. Li, and W. W. Yang, "Laboratory research on combined cooling, heating and power (CCHP) systems," *Energy Conversion and Management*, vol. 50, no. 4, pp. 977–982, 2009.

[122] M. Badami and A. Portoraro, "Performance analysis of an innovative small-scale trigeneration plant with liquid desiccant cooling system," *Energy and Buildings*, vol. 41, no. 11, pp. 1195–1204, 2009.

[123] R. Easow and P. Muley, "Micro-trigeneration:-The best way for decentralized power, cooling and heating," in *Proceedings of IEEE Conference on Innovative Technologies for an Efficient and Reliable Electricity Supply (CITRES)*, Waltham, MA, USA, September 27–29, 2010, pp. 459–466.

[124] Southeast CHP Application Center, *"NC solar center integrated micro-CHP and solar system,"* Southeast CHP Application Center, Tech. Rep., 2010.

[125] G. Angrisani, A. Rosato, C. Roselli, M. Sasso, and S. Sibilio, "Experimental results of a micro-trigeneration installation," *Applied Thermal Engineering*, vol. 38, pp. 78–90, 2012.

[126] K. K. Khatri, D. Sharma, S. L. Soni, and D. Tanwar, "Experimental investigation of CI engine operated micro-trigeneration system," *Applied Thermal Engineering*, vol. 30, no. 11-12, pp. 1505–1509, 2010.

[127] X. Q. Kong, R. Z. Wang, Y. Li, and X. H. Huang, "Optimal operation of a micro-combined cooling, heating and power system driven by a gas engine," *Energy Conversion and Management*, vol. 50, no. 3, pp. 530–538, 2009.

[128] K. Gluesenkamp, Y. Hwang, and R. Radermacher, "High efficiency micro trigeneration systems," *Applied Thermal Engineering*, vol. 50, no. 2, pp. 1480–1486, 2013.

[129] M. Ebrahimi, A. Keshavarz, and A. Jamali, "Energy and exergy analyses of a micro-steam CCHP cycle for a residential building," *Energy and Buildings*, vol. 45, pp. 202–210, 2012.

[130] A. Cervone, D. Z. Romito, and E. Santini, "Technical and economic analysis of a micro-tri/cogeneration system with reference to the primary power source in a shopping center," in *Proceedings of International Conference on Clean Electrical Power*, Ischia, June 14–16, 2011, pp. 439–445.

[131] K. Uzuneanu and D. Scarpete, "Energetic and environmental analysis of a micro CCHP system for domestic use," in *Proceedings of IASME/WSEAS International Conference on Energy & Environment*, 2011, pp. 322–327.

[132] M. Ameri, A. Behbahaninia, and A. A. Tanha, "Thermodynamic analysis of a tri-generation system based on micro-gas turbine with a steam ejector refrigeration system," *Energy*, vol. 35, no. 5, pp. 2203–2209, 2010.

[133] A. Moran, P. J. Mago, and L. M. Chamra, "Thermoeconomic modeling of micro-CHP (micro-cooling, heating, and power) for small commercial applications," *International Journal of Energy Research*, vol. 32, no. 9, pp. 808–823, 2008.

[134] J. Deng, R. Z. Wang, J. Wu, G. Han, D. Wu, and S. Li, "Exergy cost analysis of a micro-trigeneration system based on the structural theory of thermoeconomics," *Energy*, vol. 33, no. 9, pp. 1417–1426, 2008.

[135] S. Arosio, M. Guilizzoni, and F. Pravettoni, "A model for micro-trigeneration systems based on linear optimization and the Italian tariff policy," *Applied Thermal Engineering*, vol. 31, no. 14-15, pp. 2292–2300, 2011.

[136] T. J. Tracy, "Design, modeling, construction, and flow splitting optimization of a micro combined heating, cooling, and power system," Master's thesis, Florida State University, 2010.

[137] A. I. Palmero-Marrero and A. C. Oliveira, "Performance simulation of a solar-assisted micro-tri-generation system: hotel case study," *International Journal of Low-Carbon Technologies*, vol. 6, no. 4, pp. 309–317, 2011.

[138] F. Immovilli, A. Bellini, C. Bianchini, and G. Franceschini, "Solar trigeneration for residential applications, a feasible alternative to traditional microcogeneration and trigeneration plants," in *IEEE Industry Applications Society Annual Meeting*, 2008, pp. 1–8.

[139] Y. Huangfu, J. Y. Wu, R. Z. Wang, and Z. Z. Xia, "Experimental investigation of adsorption chiller for micro-scale BCHP system application," *Energy and Buildings*, vol. 39, no. 2, pp. 120–127, 2007.

[140] Northeast Clean Energy Application Center, "Cooley Dickinson–500 kW biomass CHP plant," Northeast Clean Energy Application Center, Tech. Rep., 2016.

[141] Pacific Region CHP Application Center, "East Bay Municipal Utility District–600 kW microturbine CHP/chiller system," Pacific Region CHP Application Center, Tech. Rep., 2006.

[142] Northeast CHP Application Center, "Smithfield Gardens Assisted Living Community–75 kW CHP plant," Northeast CHP Application Center, Tech. Rep., 2011.

[143] Pacific Region CHP Application Center, "Vineyard 29–120 kW microturbine/chiller system," Pacific Region CHP Application Center, Tech. Rep., 2007.

[144] M. Badami, M. Ferrero, and A. Portoraro, "Experimental assessment of a small-scale trigeneration plant with a natural gas microturbine and a liquid desiccant system," in *Proceedings of European Conference on Polygeneration*, Tarragona, Spain, March 30–April 1, 2011, pp. 1–10.

[145] P. A. Katsigiannis and D. P. Papadopoulos, "A systematic computational procedure for assessing small-scale cogeneration application schemes," in *Proceedings of International Conference on Power Engineering, Energy and Electrical Devices*, Setubal, Portugal, April 12–14, 2007, pp. 201–206.

[146] Northeast CHP Application Center, "Harbec Plastics–750 kW CHP application," Northeast CHP Application Center, Tech. Rep., 2006.

[147] Pacific Region CHP Application Center, "The Ritz-Carlton Hotel in San Francisco–240 kW Microturbine/absorption chiller system," Pacific Region CHP Application Center, Tech. Rep., 2006.

[148] R. de Boer, S. F. Smeding, and R. J. H. Grisel, "Performance of a silica-gel + water adsorption cooling system for use in small-scale tri-generation applications," in *Proceedings of Heat Powered Cycles Conference*, 2006, pp. 1–13.

[149] G. Chicco and P. Mancarella, "Planning aspects and performance indicators for small-scale trigeneration plants," in *Proceedings of International Conference on Future Power Systems*, Amsterdam, Netherlands, November 18, 2005, pp. 1–6.

[150] R. Boukhanouf, J. Godefroy, S. B. Riffat, and M. Worall, "Design and optimisation of a small-scale tri-generation system," *International Journal of Low-Carbon Technologies*, vol. 3, no. 1, pp. 32–43, 2008.

[151] L. Lin, Y. Wang, T. Al-Shemmeri, T. Ruxton, S. Turner, S. Zeng, J. Huang, Y. He, and X. Huang, "An experimental investigation of a household size trigeneration," *Applied Thermal Engineering*, vol. 27, no. 2–3, pp. 576–585, 2007.

[152] G. Abdollahia and M. Meratizaman, "Multi-objective approach in thermoenvironomic optimization of a small-scale distributed CCHP system with risk analysis," *Energy and Buildings*, vol. 43, no. 11, pp. 3144–3153, 2011.

[153] A. Canova, C. Cavallero, F. Freschi, L. Giaccone, M. Repetto, and M. Tartaglia, "Comparative economical analysis of a small scale trigenerative plant: A case study," in *Proceedings of IEEE Industry Applications Conference*, New Orleans, LA, USA, September 23–27, 2007, pp. 1456–1459.

[154] A. K. Hossain, R. Thorpe, R. E. Critoph, and P. A. Davies, "Development of a small-scale trigeneration plant based on a CI engine fuelled by neat non-edible plant oil," *Journal of Scientific & Industrial Research*, vol. 70, no. 8, pp. 688–693, 2011.

[155] Midwest CHP Application Center, "Elgin Community College–4.1 MW CHP application," Midwest CHP Application Center, Tech. Rep., 2005.

[156] Southeast CHP Application Center, "James H. Quillen VA Medical Center–3.2 MW CHP system," Southeast CHP Application Center, Tech. Rep., 2015.

[157] Midwest CHP Application Center, "Spectrum health, butterworth campus–3.8 MW CHP application," Midwest CHP Application Center, Tech. Rep., 2006.

[158] Midwest CHP Application Center, "Central Connecticut State University–2.5 MW CHP application," Midwest CHP Application Center, Tech. Rep., 2006.

[159] Northeast Clean Energy Application Center, "Bradley Airport Energy Center—5.8 MW CHP plant," Northeast Clean Energy Application Center, Tech. Rep., 2011.

[160] S. Popli, P. Rodgers, and V. Eveloy, "Trigeneration scheme for energy efficiency enhancement in a natural gas processing plant through turbine exhaust gas waste heat utilization," *Applied Energy*, vol. 93, pp. 624–636, 2012.

[161] P. A. Rodriguez-Aumente, M. del Carmen Rodriguez-Hidalgo, J. I. Nogueira, A. Lecuona, and M. del Carmen Venegas, "District heating and cooling for business buildings in Madrid," *Applied Thermal Engineering*, vol. 50, no. 2, pp. 1496–1503, 2013.

[162] O. Balli, H. Aras, and A. Hepbasli, "Thermodynamic and thermoeconomic analyses of a trigeneration (TRIGEN) system with a gas–diesel engine: Part I—methodology," *Energy Conversion and Management*, vol. 51, no. 11, pp. 2252–2259, 2010.

[163] O. Balli, H. Aras, and A. Hepbasli, "Thermodynamic and thermoeconomic analyses of a trigeneration (TRIGEN) system with a gas–diesel engine: Part II—an application," *Energy Conversion and Management*, vol. 51, no. 11, pp. 2260–2271, 2010.

[164] M. A. Meybodi and M. Behnia, "Impact of carbon tax on internal combustion engine size selection in a medium scale CHP system," *Applied Energy*, vol. 88, no. 12, pp. 5153–5163, 2011.

[165] W. L. Becker, R. J. Braun, M. Penevb, and M. Melaina, "Design and technoeconomic performance analysis of a 1 MW solid oxide fuel cell polygeneration system for combined production of heat, hydrogen, and power," *Journal of Power Sources*, vol. 200, pp. 34–44, 2012.

[166] Midwest CHP Application Center, *"University of Michigan—45.2 MW CHP application,"* Midwest CHP Application Center, Tech. Rep., 2012.

[167] Pacific Region CHP Application Center, "University of California at San Diego–30 MW CHP system," Pacific Region CHP Application Center, Tech. Rep., 2008.

[168] Midwest CHP Application Center, "University of Illinois at Chicago–57.4 MW CHP application," Midwest CHP Application Center, Tech. Rep., 2001.

[169] Southeast CHP Application Center, "UNC Chapel Hill–32 MW cogeneration plant," Southwest CHP Application Center, Tech. Rep., 2009.

[170] Gulf Coast Clean Energy Application Center, "University of Texas, Austin–137 MW (65 MW-peak) CHP application," Gulf Coast Clean Energy Application Center, Tech. Rep., 2011.

[171] Southeast CHP Application Center, "Vanderbilt University Plant Operations–25 MW cogeneration plant," Southeast CHP Application Center, Tech. Rep., 2010.

[172] P. J. Mago and L. M. Chamra, "Analysis and optimization of CCHP systems based on energy, economical, and environmental considerations," *Energy and Buildings*, vol. 41, no. 10, pp. 1099–1106, 2009.

[173] E. Cardona and A. Piacentino, "A methodology for sizing a trigeneration plant in mediterranean areas," *Applied Thermal Engineering*, vol. 23, no. 13, pp. 1665–1680, 2003.

[174] E. Cardona and A. Piacentino, "Matching economical, energetic, and environmental benefits: An analysis for hybrid CCHP-heat pump systems," *Energy*, vol. 47, no. 20, pp. 3530–3542, 2006.

[175] H. Cho, P. J. Mago, R. Luck, and L. M. Chamra, "Evaluation of CCHP systems performance based on operational cost, primary energy concumption, and carbon dioxide emission by utilizing an optimal operation scheme," *Applied Energy*, vol. 86, no. 12, pp. 2540–2549, 2009.

[176] P. J. Mago, N. Fumo, and L. M. Chamra, "Performance analysis of CCHP and CHP systems operating following the thermal and electric load," *International Journal of Energy Research*, vol. 33, no. 9, pp. 852–864, 2009.

[177] J. Wang, C. Zhang, and Y. Jing, "Multi-criteria analysis of combined cooling, heating and power systems in different climate zones in China," *Applied Energy*, vol. 87, no. 4, pp. 1247–1259, 2010.

[178] A. Smith, R. Luck, and P. J. Mago, "Analysis of a combined cooling, heating, and power system model under different operating strategies with input and model data uncertainty," *Energy and Buildings*, vol. 42, no. 11, pp. 2231–2240, 2010.

[179] J. Wang, Y. Jing, C. Zhang, and Z. Zhai, "Performance comparison of combined cooling heating and power system in different operation modes," *Applied Energy*, vol. 88, no. 12, pp. 4621–4631, 2011.

[180] N. Fumo, P. J. Mago, and A. D. Smith, "Analysis of combined cooling, heating, and power systems operating following the electric load and following the thermal load strategies with no electricity export," *Proceedings of the Institution of Mechanical Engineers, Part A: Journal of Power and Energy*, vol. 225, pp. 1016–1025, 2011.

[181] M. Liu, Y. Shi, and F. Fang, "A new operation strategy for CCHP systems with hybrid chillers," *Applied Energy*, vol. 95, pp. 164–173, 2012.

[182] R. Hashemi, "A developed offline model for optimal operation of combined heating and cooling and power systems," *IEEE Transactions on Energy Conversion*, vol. 24, no. 1, pp. 222–229, 2009.

[183] N. Fumo and L. M. Chamra, "Analysis of combined cooling, heating, and power systems based on source primary energy consumption," *Applied Energy*, vol. 87, no. 6, pp. 2023–2030, 2010.

[184] E. Cardona, A. Piacentino, and F. Cardona, "Energy saving in airports by trigeneration. Part II: Short and long term planning for the Malpensa 2000 CHCP plant," *Applied Thermal Engineering*, vol. 26, no. 14-15, pp. 1437–1447, 2006.

[185] N. Fumo, P. J. Mago, and L. M. Chamra, "Emission operational strategy for combined cooling, heating, and power systems," *Applied Energy*, vol. 86, no. 11, pp. 2344–2350, 2009.

[186] F. Fang, Q. H. Wang, and Y. Shi, "A novel optimal operational strategy for the CCHP system based on two operating modes," *IEEE Transactions on Power Systems*, vol. 27, no. 2, pp. 1032–1041, 2011.

[187] A. Zafra-Cabeza, M. A. Ridao, I. Alvarado, and E. F. Camacho, "Applying risk management to combined heat and power plants," *IEEE Transactions on Power Systems*, vol. 23, no. 3, pp. 938–945, 2008.

[188] M. A. Lozano, M. Carvalho, and L. M. Serra, "Operational strategy and marginal costs in simple trigeneration systems," *Energy*, vol. 34, no. 11, pp. 2001–2008, 2009.

[189] A. Nosrat and J. M. Pearce, "Dispatch strategy and model for hybrid photovoltaic and trigeneration power systems," *Applied Energy*, vol. 88, no. 9, pp. 3270–3276, 2011.

[190] C. Rubio-Maya, J. Uche-Marcuello, A. Martínez-Gracia, and A. A. Bayod-Rújula, "Design optimization of a polygeneration plant fuelled by natural gas and renewable energy sources," *Applied Energy*, vol. 88, no. 2, pp. 449–457, 2011.

[191] C. Rubio-Maya, J. Uche, and A. Martínez, "Sequential optimization of a polygeneration plant," *Energy Conversion and Management*, vol. 52, no. 8-9, pp. 2861–2869, 2011.

[192] D. Buoro, M. Casisi, P. Pinamonti, and M. Reini, "Optimization of distributed trigeneration systems integrated with heating and cooling micro-grids," *Distributed Generation & Alternative Energy Journal*, vol. 26, no. 2, pp. 7–34, 2011.

[193] C. Z. Li, Y. M. Shi, and X. H. Huang, "Sensitivity analysis of energy demands on performance of CCHP system," *Energy Conversion and Management*, vol. 49, no. 12, pp. 3491–3497, 2008.

[194] M. A. Lozano, J. C. Ramos, and L. M. Serra, "Cost optimization of the design of CHCP (combined heat, cooling and power) systems under legal constraints," *Energy*, vol. 35, no. 2, pp. 794–805, 2010.

[195] C.-J. F. Tuula Savola and, "MINLP optimisation model for increased power production in small-scale CHP plants," *Applied Thermal Engineering*, vol. 27, no. 1, pp. 89–99, 2007.

[196] H. Li, R. Nalim, and P. A. Haldi, "Thermal-economic optimization of a distributed multi-generation energy system—a case study of Beijing," *Applied Thermal Engineering*, vol. 26, no. 7, pp. 709–719, 2006.

[197] A. Rong and R. Lahdelma, "An efficient linear programming model and optimization algorithm for trigeneration," *Applied Energy*, vol. 82, no. 1, pp. 40–63, 2005.

[198] Q. H. Wang and F. Fang, "Optimal configuration of CCHP system based on energy, economical, and environmental considerations," in *Proceedings of International Conference on Intelligent Control and Information Processing*, Harbin, China, July 25–28, 2011, pp. 489–494.

[199] A. Sheikhi, B. Mozafari, and A. M. Ranjbar, "CHP optimized selection methodology for a multi-carrier energy system," in *IEEE Trondheim PowerTech*, 2011, pp. 1–7.

[200] K. C. Kavvadias and Z. B. Maroulis, "Multi-objective optimization of a trigeneration plant," *Energy Policy*, vol. 38, no. 2, pp. 945–954, 2010.

[201] J. Wang, Z. Zhai, Y. Jing, and C. Zhang, "Optimization design of BCHP system to maximize to save energy and reduce environmental impact," *Energy*, vol. 35, no. 8, pp. 3388–3398, 2010.

[202] J. Wang, Y. Jing, C. Zhang, G. Shi, and X. Zhang, "A fuzzy multi-criteria decision-making model for trigeneration system," *Energy Policy*, vol. 36, no. 10, pp. 3823–3832, 2008.

[203] A. Piacentino and F. Cardona, "EABOT–energetic analysis as a basis for robust optimization of trigeneration systems by linear programming," *Energy Conversion and Management*, vol. 49, no. 11, pp. 3006–3016, 2008.

[204] G. Chicco and P. Mancarella, "Matrix modelling of small-scale trigeneration systems and application to operational optimization," *Energy*, vol. 34, no. 3, pp. 261–273, 2009.

[205] M. Geidl and G. Andersson, "Optimal power dispatch and conversion in systems with multiple energy carriers," in *Proceedings of Power Systems Computation Conference*, 2005, pp. 1–7.

[206] M. Geidl and G. Andersson, "Optimal power flow of multiple energy carriers," *IEEE Transactions on Power Systems*, vol. 22, no. 1, pp. 145–155, 2007.

[207] M. Geidl, G. Koeppel, P. Favre-Perrod, B. Klockl, G. Andersson, and K. Frohlich, "Energy hubs for the future," *IEEE Power & Energy Magzine*, vol. 5, no. 1, pp. 24–30, 2007.

[208] M. Geidl and G. Andersson, "A modeling and optimization approach for multiple energy carrier power flow," in *Proceedings of IEEE Russia Power Tech Conference*, St. Petersburg, Russia, June 27–30, 2005, pp. 1–7.

[209] H. Ghaebi, M. H. Saidi, and P. Ahmadi, "Exergoeconomic optimization of a trigeneration system for heating, cooling and power production purpose based on TRR method and using evolutionary algorithm," *Applied Thermal Engineering*, vol. 36, pp. 113–125, 2012.

[210] A. K. Hueffed and P. J. Mago, "Influence of prime mover size and operational strategy on the performance of combined cooling, heating, and power systems under different cost structures," *Proceedings of the Institution of Mechanical Engineers, Part A: Journal of Power and Energy*, vol. 224, pp. 591–605, 2010.

[211] A. Sclafani and A. Beyene, "Sizing CCHP systems for variable and non-coincident loads: Part 1—load profiling and equipment selection," *Cogeneration and Distributed Generation Journal*, vol. 23, no. 3, pp. 6–19, 2008.

[212] O. A. Shaneb, G. Coates, and P. C. Taylor, "Sizing of residential $\mu$CHP systems," *Energy and Buildings*, vol. 43, no. 8, pp. 1991–2001, 2011.

[213] J. Harrod, P. J. Mago, and R. Luck, "Sizing analysis of a combined cooling, heating, and power system for a small office building using a wood waste biomass-fired stirling engine," *International Journal of Energy Research*, vol. 36, no. 1, pp. 64–74, 2010.

[214] H. Ren, W. Gao, and Y. Ruan, "Optimal sizing for residential CHP system," *Applied Thermal Engineering*, vol. 28, no. 5–6, pp. 514–523, 2008.

[215] B. Zhang and W. Long, "An optimal sizing method for cogeneration plants," *Energy and Buildings*, vol. 38, no. 3, pp. 189–195, 2006.

[216] A. H. Azit and K. M. Nor, "Optimal sizing for a gas-fired grid-connected cogeneration system planning," *IEEE Transactions on Energy Conversion*, vol. 24, no. 4, pp. 950–958, 2009.

[217] C. Rubio-Maya, J. Uche, and A. Martínez-Gracia, "Selection and sizing procedure of polygeneration plants using mathematical programming," in *Proceedings of International Conference on Efficiency, Cost, Optimization, Simulation & Environmental Impact of Energy Systems*, 2009, pp. 587–594.

[218] S. Martínez-Lera and J. Ballester, "A novel method for the design of CHCP (combined heat, cooling and power) systems for buildings," *Energy*, vol. 35, no. 7, pp. 2972–2984, 2010.

[219] A. Sheikhi, A. M. Ranjbar, and H. Oraee, "Financial analysis and optimal size and operation for a multicarrier energy system," *Energy and Buildings*, vol. 48, pp. 71–78, 2011.

[220] B. Hedman, "CHP: The state of the market," in *Proceedings of U.S. EPA Combined Heat and Power Partnership Partners Meeting & NYSERDA CHP Roundtable*, 2009, pp. 1–51.

[221] International District Energy Association, *Combined Heat and Power (CHP): Essential for a Cost Effective Clean Energy Standard*. The International District Energy Association (IDEA), 2011.

[222] PEW Center on Global Climate Change, "Cogeneration/Combined heat and power (CHP)," PEW Center on Global Climate Change, Tech. Rep., 2011.

[223] I. MacLeay, K. Harris, A. Annut, and chapter authors, "Digest of United Kingdom energy statistics 2011," UK Department of Energy & Climate Change, Tech. Rep., 2011.

[224] UK Department of Energy & Climate Change, "Planning our electric future: A White Paper for secure, affordable and lowcarbon electricity," UK Department of Energy & Climate Change, Tech. Rep., 2011.

[225] T. Kerr, "CHP and DHC in China: An assessment of market and policy potential," International Energy Agency, Tech. Rep., 2008.

# 2

# An Optimal Switching Strategy for Operating CCHP Systems

## 2.1 Introduction and Related Work

The commonly used operation strategies for CCHP systems are FEL and FTL [1]. At any time, one of them can be selected as the current strategy based on the performance criteria and constraints. In order to obtain an effective operation strategy, two primary steps are necessary: the construction of the performance criteria and the optimal design of the strategy.

The performance criteria play a decisive role in the selection of the current operating mode. Similar to many other energy conversion and supply systems, the primary purpose of the CCHP system is to satisfy the energy requirements of users. Accordingly, the simplest criterion is the expression of the supply–demand balance of energy [1, 2]. Unfortunately, all energy supply systems no longer serve a single target. Some critical factors, such as energy conversion efficiency and environmental impacts, are required to be taken into account simultaneously. From this perspective, PEC/PES [3, 4], CDE [5–7], seasonal atmospheric conditions [8], and others are added to the criteria. Furthermore, some researchers have integrated multiple factors into one criterion to improve the overall performance of the CCHP system [9, 10]. For such studies, setting and tuning the weight of each impact factor are vital and need further investigations.

Based on certain performance criteria, some researchers such as Cardona, Zafra-Cabeza, Mago, Smith, Chicco, among others, investigated the optimal operation strategies of the CCHP system. Cardona *et al.* [11] assessed the technical and economic feasibility of the CCHP system in airports based on TDM and EDM. They presented a profit-oriented optimal management strategy by taking into account the articulated energy tariff system and the technical characteristics of components [12]. Zafra-Cabeza *et al.* [13] presented a risk management strategy for load schedule with uncertainties and multi-constraints. They adopted the MPC method to determine the

*Combined Cooling, Heating, and Power Systems: Modeling, Optimization, and Operation*, First Edition.
Yang Shi, Mingxi Liu, and Fang Fang.
© 2017 John Wiley & Sons Ltd. Published 2017 by John Wiley & Sons Ltd.

risk mitigation actions for lower impacts from possible risks. Mago and Chamra [14] proposed HETS, which is a good alternative for the operation of the CCHP system since it can yield good reductions of PEC, COST, and CDE. Smith *et al.* [15] characterized the uncertainties in a representative steady-state model and input parameters of the CCHP system such as the thermal load, natural gas and electricity prices, and engine performance. They presented a complete performance evaluation in terms of sensitivity and uncertainty analysis. Chicco and Mancarella [16] illustrated and evaluated the possible benefits of adopting different trigeneration alternatives in the optimization of a CCHP system, specifically focusing on comparing six alternative cooling production solutions. Their work provided a useful framework for carrying out trigeneration analyses in various contexts.

For the optimal operation problem of the CCHP system, the performance criteria can be generally regarded as objective functions. To achieve an expected optimal objective, some practical constraints according to the equipment status and running conditions, and the corresponding optimization algorithms should be developed for the CCHP system. Rong and Lahdelma [17] modeled the hourly trigeneration problem as an LP model to minimize the production and purchase costs of three energy components, as well as $CO_2$ emissions costs. They proposed a specialized tri-commodity simplex (TCS) algorithm to solve the optimization problem. Piacentino and Cardona [18] paid attention to the energetic analysis of the CCHP system with economical and energo-environmental implications. Using the LP interior point method, they developed an energy-analysis-based optimization of trigeneration plants (EABOT) tool for the CCHP system including thermal energy storages. Wang *et al.* [19] employed the particle swarm optimization algorithm (PSOA) to the design and operation of the CCHP system. They formulated a PSOA-based optimization procedure considering the performance characteristics of equipment, the energy flow relationship, and the operating constraints.

This chapter focuses on the optimization of the operation strategy for the existing CCHP system. The motivation of this chapter is to take full advantage of various performance criteria and to improve the comprehensive performance of the CCHP system. For this purpose, an optimal switching operation strategy is presented. Using this strategy, the whole operating space of the CCHP system is divided into several regions by two or three border surfaces, which are decided by an integrated performance criterion, that is, EC. The operating point of the CCHP system is scheduled to the corresponding region to switch between FEL and FTL. Furthermore, a reference CCHP system modeled for a hypothetical hotel in Beijing, China, is used to assess the effectiveness and performance of the proposed strategy.

The remainder of this chapter is organized as follows: Section 2.2 elaborates FEL and FTL for the CCHP system. Section 2.3 formulates the EC function, which includes the PEC, CDE, and COST. Section 2.4 presents the EC-based optimal switching operation strategy. Section 2.5 demonstrates the case studies based on a hypothetical CCHP system. Section 2.6 summarizes this chapter.

## 2.2 Conventional Operation Strategies of CCHP Systems

The energy flow of an SP system is shown in Figure 2.1. An electric chiller is adopted in the cooling system. The heat, coming from a natural gas boiler, is distributed through heating units. The electricity, needed by users and the electric chiller, comes from the power grid.

For the SP system, the total electricity consumption from the power grid is

$$E_{grid}^{SP} = E_{user} + E_{ec}^{SP}. \tag{2.1}$$

The static relationship between the input and the output of the electric chiller can be expressed as

$$E_{ec}^{SP} = \frac{Q_c}{COP_{ec}}. \tag{2.2}$$

The fuel consumption of the heating system is defined as

$$F_b^{SP} = \frac{Q_b^{SP}}{\eta_b} = \frac{Q_h}{\eta_b \eta_h}. \tag{2.3}$$

The main distinction between the CCHP system and the SP system lies in that the waste heat rejected from the CCHP's prime mover is recovered to produce cooling and heating to meet the needs of facilities. There are several kinds of typical configurations for the CCHP system [20]. In this chapter, the reference CCHP system includes a PGU with a gas turbine, a heat recovery system, an auxiliary boiler, an absorption chiller, and a heating unit. Its schematic is shown in Figure 2.2. When there is not enough recovered heat to handle the thermal (cooling and heating) requirements, the auxiliary boiler will be used to compensate for the shortfall.

As mentioned before, the CCHP system has two typical operating modes: FEL and FTL. After the structure and devices have been finalized, the efficiency and performance then depend largely on the selection of operating modes.

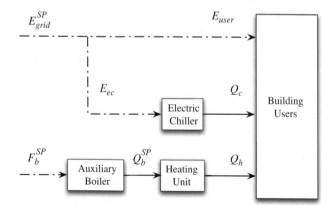

**Figure 2.1** Schematic of a typical SP system

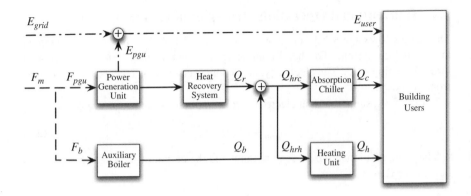

**Figure 2.2**    Schematic of a typical CCHP system

To facilitate the following analysis and synthesis, some essential assumptions are made as follows:

(S1) The structure and size of the CCHP system have been fixed and the rated capacity (including the auxiliary boiler) is sufficient to satisfy the peak load of electricity and thermal energy.

(S2) The minimum technical limit of the CCHP system is neglected. All the involved equipment can operate ranging from 0 to 100% of its rated capacity.

(S3) The efficiency drops of CCHP equipment at partial load operation are neglected to simplify the analysis and calculation.

(S4) The electric power consumption of pumps, fans and other distribution equipment and the transmission loss of grid are not considered.

Admittedly, Assumptions (S2) and (S3) are relatively stringent. In practice, few of the equipment can work steadily at a very low load, and their efficiencies always decrease with the load reduction. In this case, the assumption-based optimization results may not fully reflect the actual situation. Nevertheless, partial load and unsteady operation of the CCHP system are very complicated issues. Although some results have been achieved, for example [21, 22], the relevant research on partial load of the CCHP system is still challenging due to its inherent complexity. Therefore, to focus more on the development of operation strategy design, and to make the obtained results practical and feasible, we adopt these assumptions.

## 2.2.1   FEL Mode of the CCHP System

FEL mode means that the PGU is loaded to satisfy the electric requirement of users. At the same time, the waste heat from this loading is recovered to supply the cooling and heating requirements [1]. Unfortunately, in most cases, the recovered heat cannot exactly satisfy the thermal requirement. Therefore, the auxiliary boiler and, under certain circumstances, heat storage equipment should be implemented in the CCHP system to balance the energy supply–demand.

First, the total electricity generated by PGU must satisfy the requirement of users, indicating

$$E_{pgu-FEL} = E_{user}. \tag{2.4}$$

The fuel consumption of PGU can be estimated as

$$F_{pgu-FEL} = \frac{E_{pgu-FEL}}{\eta_{pgu}}. \tag{2.5}$$

Consequently, the thermal energy transformed by the heat recovery system is given by

$$\begin{aligned} Q_{r-FEL} &= (F_{pgu-FEL} - E_{pgu-FEL})\eta_{hrs} \\ &= F_{pgu-FEL}(1 - \eta_{pgu})\eta_{hrs}. \end{aligned} \tag{2.6}$$

$Q_{r-FEL}$ could be either more or less than (or equal to) users' requirements, yielding the following algorithm. Herein, $Q_{hrc} = Q_c/COP_{ac}$, $Q_{hrh} = Q_h/\eta_h$.

---

**Algorithm**  FEL mode

**procedure**
  **if** $Q_{r-FEL} \le Q_{hrc} + Q_{hrh}$ **then**
    $Q_{b-FEL} = Q_{hrc} + Q_{hrh} - Q_{r-FEL}$
  **else**
    $Q_{b-FEL} = 0$
  **end if**
**end procedure**

---

Then, the fuel consumption of the auxiliary boiler can be calculated as

$$F_{b-FEL} = \frac{Q_{b-FEL}}{\eta_b}. \tag{2.7}$$

Consequently, the total fuel consumption that is registered at the meter can be estimated as

$$F_{m-FEL} = F_{pgu-FEL} + F_{b-FEL}. \tag{2.8}$$

## 2.2.2 FTL Mode of the CCHP System

FTL mode means that the PGU is loaded to have adequate recovered waste heat to supply the cooling and heating requirements. Meanwhile, if the produced electricity is not equal to the request, the discrepancies have to be imported from or sent back to the grid.

First, the total thermal energy, generated by the heat recovery system, must match the thermal energy requirement from the absorption chiller and heating unit, indicating

$$Q_{r-FTL} = Q_{hrc} + Q_{hrh}. \tag{2.9}$$

Since the recovered thermal energy from the prime mover is known, the fuel consumption can be estimated as

$$F_{pgu-FTL} = \frac{Q_{r-FTL}}{\eta_{hrs}(1 - \eta_{pgu})}. \tag{2.10}$$

Next, the total electric power supplied by the PGU can be determined as

$$E_{pgu-FTL} = F_{pgu-FTL}\eta_{pgu}. \tag{2.11}$$

The generated electric power may be either more or less than (or equal to) users' requirements, yielding the following algorithm.

---

**Algorithm**    FTL mode

---

  **procedure**
    **if** $E_{pgu-FTL} \leq E_{user}$ **then**
      $E_{grid} = E_{user} - E_{pgu-FTL}$
    **else**
      $E_{excess} = E_{pgu-FTL} - E_{user}$
    **end if**
  **end procedure**

---

## 2.3    EC Function and the Optimal Switching Operation Strategy

In order to achieve required performance, some criteria should be formulated as the decision basis to determine the operating mode of the CCHP system [10]. Here, we take the CCHP system as an independent production unit to calculate its absolute energy consumption and resulting emissions.

### 2.3.1    PEC

PEC of the CCHP system is the most common factor to be considered for the selection of operating modes. The PEC criteria for FEL mode and FTL mode can be expressed as

$$PEC_{FEL}^{CCHP} = F_{m-FEL}k_f, \tag{2.12}$$

$$PEC_{FTL}^{CCHP} = E_{grid}k_e + F_{m-FTL}k_f. \tag{2.13}$$

If $PEC_{FEL}^{CCHP} < PEC_{FTL}^{CCHP}$, FEL mode should be selected as the current operating mode for less PEC; otherwise FTL mode should be adopted. Sometimes $PEC_{FEL}^{CCHP} = PEC_{FTL}^{CCHP}$, which means that the consumption of primary energy is equivalent whenever in FEL mode or FTL mode.

### 2.3.2 CDE

Mounting scientific evidence shows that the $CO_2$ emitted by fossil-fuel combustion contributes to global warming. Therefore, reducing $CO_2$ emissions of the CCHP system is an important issue; the CDE criteria for FEL and FTL mode are given by

$$CDE_{FEL}^{CCHP} = F_{m-FEL}\mu_f, \tag{2.14}$$

$$CDE_{FTL}^{CCHP} = E_{grid}\mu_e + F_{m-FTL}\mu_f. \tag{2.15}$$

As mentioned before, the operating mode with a smaller CDE criterion should be selected.

### 2.3.3 COST

Five Nordic countries, Denmark, Finland, Norway, Netherlands, and Sweden, have implemented a carbon tax or energy tax. Therefore, the electric power costs, fuel costs, and carbon tax can all be included in the COST criterion as

$$COST_{FEL}^{CCHP} = F_{m-FEL}C_f + F_{m-FEL}\mu_f C_{ca}, \tag{2.16}$$

$$COST_{FTL}^{CCHP} = E_{grid}C_e + F_{m-FTL}C_f + F_{m-FTL}\mu_f C_{ca} - E_{excess}C_s. \tag{2.17}$$

Taking into account that the possible excess electricity produced in FTL mode can be sold back to the grid, and sales revenue can be used to offset the COST, we add $-E_{excess}C_s$ into (2.17). But if sales are not allowed, we can set $C_s = 0$ to omit this item.

### 2.3.4 EC Function

Each of the aforementioned three criteria can be used separately to determine the operating mode of the CCHP system. However, at the same time, according to different criteria, the result of the mode selection may be different, bringing confusion to the system operator. Therefore, it is necessary to formulate an EC to get a comprehensive and unique optimal result.

For a normalized value of EC, a contrast system should be chosen first. As described in Section 2.2, the SP system can provide users with cold, heat, and electricity in separate ways, and its efficiency and performance are usually lower than that of the CCHP system. Furthermore, the PEC, CDE, and COST criteria of the SP system can also be formulated by referring to the CCHP system. For these reasons, the SP system is selected as the contrast system in this chapter.

The SP system supplies cooling and heating to users by converting electricity and fuel. Its PEC value is given by

$$PEC^{SP} = E^{SP}_{grid}k_e + F^{SP}_b k_f. \tag{2.18}$$

Under normal circumstances, the conversion factors $k_e$ and $k_f$ are based on two separate quantities: (1) the quantity of source energy for fuel and electricity generation; and (2) the quantity of fuel and electricity lost in transmission and distribution. Suppose that the SP and CCHP systems are located in the same or an adjacent area, then $k_e$ is the same for both systems, and so is $k_f$ .

The CDE value of the SP system can be calculated as

$$CDE^{SP} = E^{SP}_{grid}\mu_e + F^{SP}_b \mu_f. \tag{2.19}$$

Emission conversion factors of the SP and CCHP systems can also be considered as equal.

The COST value of the SP system is derived as

$$COST^{SP} = E^{SP}_{grid}C_e + F^{SP}_b C_f + F^{SP}_b \mu_f C_{ca}. \tag{2.20}$$

In the same region, energy prices or carbon tax are the same for the SP and CCHP systems during a certain period.

Using (2.12)–(2.20), the EC value can be obtained as

$$EC = \omega_1 \frac{PEC^{CCHP}}{PEC^{SP}} + \omega_2 \frac{CDE^{CCHP}}{CDE^{SP}} + \omega_3 \frac{COST^{CCHP}}{COST^{SP}}, \tag{2.21}$$

where $\omega_1$, $\omega_2$, and $\omega_3$ are the weights of the ratios of PEC, CDE, and COST, respectively. The boundary conditions are $0 \leq \omega_1, \omega_2, \omega_3 \leq 1$, and $\omega_1 + \omega_2 + \omega_3 = 1$.

These three weights can be assigned to reflect the relative importance of each sub-criterion, and they will directly affect the optimization results. Note that, the equal-weighting method is applied in many decision-making problems as Dawes and Corrigan [23] argued that the equal-weighting method can, in most circumstances, produce results nearly as good as those for unequal-weighting methods.

### 2.3.5   Optimal Switching Operation Strategy

It is evident that the lower EC, the better performance of the CCHP system. Based on (2.21), $EC_{FEL}$ and $EC_{FTL}$ can be obtained for two typical operating modes, respectively:

$$EC_{FEL} = \frac{1}{3} \left( \frac{PEC^{CCHP}_{FEL}}{PEC^{SP}} + \frac{CDE^{CCHP}_{FEL}}{CDE^{SP}} + \frac{COST^{CCHP}_{FEL}}{COST^{SP}} \right), \tag{2.22}$$

$$EC_{FTL} = \frac{1}{3} \left( \frac{PEC^{CCHP}_{FTL}}{PEC^{SP}} + \frac{CDE^{CCHP}_{FTL}}{CDE^{SP}} + \frac{COST^{CCHP}_{FTL}}{COST^{SP}} \right), \tag{2.23}$$

where $\omega_1 = \omega_2 = \omega_3 = 1/3$. According to (2.22) and (2.23), we propose the optimal switching operation strategy as the following algorithm.

---

**Algorithm**   Optimal switching operation strategy

**procedure**
  **if** $EC_{FEL} \geq EC_{FTL}$ **then**
      Choose FTL as the current strategy
  **else**
      Choose FEL as the current strategy
  **end if**
**end procedure**

---

## 2.4   Analysis and Discussion

Combining (2.5) and (2.6), or (2.10) and (2.11), the relation between $E_{pgu}$ and $Q_r$ in both operating modes can be rewritten as

$$E_{pgu} = \frac{\eta_{pgu}}{\eta_{hrs}(1 - \eta_{pgu})}Q_r = KQ_r, \tag{2.24}$$

where $K$ represents the power to heat ratio of the CCHP system. According to (2.24), the electricity $E_{pgu}$ generated by PGU can be approximated as a linear function of the recovered heat $Q_r$.

Generally, the condition for the maximum efficiency of the CCHP system is that the power to heat ratio of energy requests should be equal to the power to heat ratio of the CCHP system (i.e., $E_{user} = KQ_{user}$). In this case, we call the electric and thermal requests of the CCHP system as balanced or matched. However, this situation rarely occurs. Therefore, it is necessary to analyze the switching conditions of operating modes and their performance based on the proposed strategy when electric and thermal requirements are not matched.

### 2.4.1   Case 1: $E_{user} \geq KQ_{user}$

The thermal request of users can be defined as

$$Q_{user} = Q_{hrc} + Q_{hrh}. \tag{2.25}$$

From the definition of $Q_{hrc}$ and $Q_{hrh}$, (2.25) can be rewritten as

$$Q_{user} = \frac{Q_c}{COP_{ac}} + \frac{Q_h}{\eta_h}. \tag{2.26}$$

In this case, if FEL mode is selected as the current operating mode, the excess thermal energy will be generated; if FTL mode is selected, the insufficient electricity will be provided by the power grid. In order to compare $EC_{FEL}$ with $EC_{FTL}$, the following relation is formulated

$$EC_{FEL} - EC_{FTL} = \frac{1}{3} \left( \frac{PEC_{FEL}^{CCHP} - PEC_{FTL}^{CCHP}}{PEC^{SP}} \right.$$

$$+ \frac{CDE_{FEL}^{CCHP} - CDE_{FTL}^{CCHP}}{CDE^{SP}} \qquad (2.27)$$

$$\left. + \frac{COST_{FEL}^{CCHP} - COST_{FTL}^{CCHP}}{COST^{SP}} \right).$$

To simplify (2.27), the differences of PEC, CDE, and COST in two operating modes should be calculated separately. The energy consumption part in (2.27) can be expressed as

$$PEC_{FEL}^{CCHP} - PEC_{FTL}^{CCHP}$$

$$= F_{m-FEL}k_f - (E_{grid}k_e + F_{m-FEL}k_f) \qquad (2.28)$$

$$= (F_{pgu-FEL} + F_{b-FEL} - F_{pgu-FTL})k_f - E_{grid}k_e.$$

When $E_{user} \geq KQ_{user}$, the thermal energy provided by the CCHP system is not less than needed, and no excess electricity is generated, that is, $F_{b-FEL} = 0$ and $E_{excess} = 0$. By substituting (2.5) and (2.11), we obtain

$$PEC_{FEL}^{CCHP} - PEC_{FTL}^{CCHP} = \left( \frac{E_{pgu-FEL}}{\eta_{pgu}} - \frac{E_{pgu-FTL}}{\eta_{pgu}} \right) k_f - E_{grid}k_e. \qquad (2.29)$$

It is obvious that $E_{pgu-FEL} = E_{user}$ in FEL mode. Consequently, the final form of (2.28) is equivalent to

$$PEC_{FEL}^{CCHP} - PEC_{FTL}^{CCHP} = E_{grid} \left( \frac{1}{\eta_{pgu}}k_f - k_e \right). \qquad (2.30)$$

According to (2.28)–(2.30), and with (2.14) and (2.15), the second part in parentheses of (2.27) is expressed as

$$CDE_{FEL}^{CCHP} - CDE_{FTL}^{CCHP} = E_{grid} \left( \frac{1}{\eta_{pgu}}\mu_f - \mu_e \right). \qquad (2.31)$$

Similarly, with (2.16) and (2.17), the third part in parentheses of (2.27) can be obtained as

$$COST_{FEL}^{CCHP} - COST_{FTL}^{CCHP} = E_{grid} \left[ \frac{C_f + \mu_f C_{ca}}{\eta_{pgu}} - C_e \right]. \qquad (2.32)$$

Substituting (2.30)–(2.32) into (2.27), we obtain

$$EC_{FEL} - EC_{FTL} = \frac{E_{grid}}{3} \left[ \frac{\eta_f - \eta_{pgu}k_e}{\eta_{pgu}PEC^{SP}} + \frac{\mu_f - \eta_{pgu}\mu_e}{\eta_{pgu}CDE^{SP}} + \frac{C_f + \mu_f C_{ca} - \eta_{pgu}C_e}{\eta_{pgu}COST^{SP}} \right].$$

(2.33)

## 2.4.2 Case 2: $E_{user} < KQ_{user}$

In this case, if FEL mode is selected as the current operating mode, the auxiliary boiler will be started up to supplement the lack of thermal energy; if FTL mode is selected, then excess electricity will be generated. The difference between $EC_{FEL}$ and $EC_{FTL}$ can also be calculated by (2.27). However, due to the change in the ratio of thermal energy and electricity, the simplification process of (2.27) and the result will be different.

When $E_{user} < KQ_{user}$, $F_{b-FEL} \neq 0$, and $E_{grid} = 0$ in (2.28), with (2.6) and (2.10), we have

$$PEC_{FEL}^{CCHP} - PEC_{FTL}^{CCHP} = \left[ \frac{Q_{r-FEL}}{(1 - \eta_{pgu})\eta_{hrs}} + F_{b-FEL} - \frac{Q_{r-FTL}}{(1 - \eta_{pgu})\eta_{hrs}} \right] k_f. \quad (2.34)$$

In FTL mode, the thermal energy produced by the CCHP system is equal to the requirement of users, that is, $Q_{r-FTL} = Q_{user} = Q_{hrc} + Q_{hrh}$. Thus, the simplified form of (2.28) is

$$PEC_{FEL}^{CCHP} - PEC_{FTL}^{CCHP} = F_{b-FEL} \left[ 1 - \frac{\eta_b}{(1 - \eta_{pgu})\eta_{hrs}} \right] k_f. \quad (2.35)$$

Following similar steps, the other parts of (2.27) can also be derived. Then, a new difference between $EC_{FEL}$ and $EC_{FTL}$ in Case 2 can be expressed as

$$EC_{FEL} - EC_{FTL} = \frac{1}{3}F_{b-FEL} \left[ 1 - \frac{\eta_b}{\eta_{hrs}(1 - \eta_{pgu})} \right]$$
$$\times \left( \frac{k_f}{PEC^{SP}} + \frac{\mu_f}{CDE^{SP}} + \frac{C_f + \mu_f C_{ca}}{COST^{SP}} \right) + \frac{1}{3} \frac{E_{excess} C_s}{COST^{SP}}. \quad (2.36)$$

## 2.4.3 Border Surfaces and Partition of Operating Modes

With the changes of energy requirements, all possible operating points of the CCHP system form a three-dimensional (3-D) space. The three directions among the space represent cooling, heating, and power, respectively. Separated by the two cases, $E_{user} \geq KQ_{user}$ and $E_{user} < KQ_{user}$, the 3-D operating space can be partitioned into two regions with a border surface (Equal-Load Interface), which is decided by $E_{pgu} = KQ_r = K(Q_{hrc} + Q_{hrh})$. Moreover, if the solutions exist by setting (2.33) and

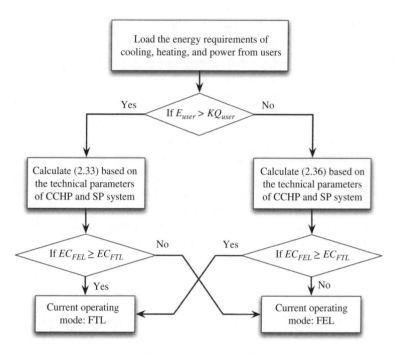

**Figure 2.3** Flow chart of the decision-making process of the proposed optimal switching operation strategy for the CCHP system based on two operating modes

(2.36) as 0, there may exist another two border surfaces (Equal-Mode Interface), which locate separately in the two partitioned operating space as the interface of FEL and FTL operating modes. Therefore, after the two-step separation, the whole 3-D operating space will be divided into two to four regions. To illustrate this more clearly, case studies about border surfaces and operating modes partition will be given in the following section.

A flow chart of the decision-making process by the proposed optimal operation strategy is shown in Figure 2.3. In practice, we can calculate the Equal-Load Interface and Equal-Mode Interface(s) in advance with the technical parameters of the CCHP system to obtain a division of the operating space. And then, the operating mode can be determined only by locating the current energy requirements in the corresponding area. With this approach, the computational load of this process will be reduced.

## 2.5   Case Study

### 2.5.1   Hypothetical Building Configuration

In this section, a CCHP system is modeled to evaluate the proposed optimal switching operation strategy. This system belongs to a hypothetical hotel in Beijing, China. It is assumed that the hotel, opening all year around, has three floors with a total construction area of 4050 m². The first floor consists of dining halls (300 m²) and

**Table 2.1**  Primary parameters of the hypothetical hotel using EnergyPlus

| Parameter | Value |
|---|---|
| Orientation | Aligned with North |
| Each floor area | 30 m×45 m |
| Each floor height | 3.5 m |
| Glass area | 40% in each wall |
| Glazing heat transfer coefficient | 4.200 W/(m²K) |
| Exterior wall heat transfer coefficient | 0.438 W/(m²K) |
| Interior wall heat transfer coefficient | 0.720 W/(m²K) |
| Floor heat transfer coefficient | 2.929 W/(m²K) |
| Roof heat transfer coefficient | 0.372 W/(m²K) |
| Electric equipment, lights and people densities | According to the public building energy-saving design standard |

office rooms (1050 m²), and the second and third floors are guest rooms. The hourly energy requirements in some representative days are estimated by EnergyPlus [24].

The primary parameters of the hypothetical hotel are listed in Table 2.1. The hourly cooling, heating and power loads in representative days of spring/autumn, summer, and winter are shown in Figure 2.4.

## 2.5.2  Test Results

The technical parameters of the CCHP system for the hypothetical hotel are listed in Table 2.2. To calculate the ECs in different operating modes, a set of necessary parameters of a reference SP system are also listed in this table. In addition, to simplify the analysis and calculation, the efficiency-related parameters of the PGU and the auxiliary boiler are fixed.

To better illustrate the characteristics of the proposed strategy, a comparison strategy HETS presented by Mago and Chamra [14] is introduced into the simulation tests. Based on (2.24), HETS can determine the operating mode of the CCHP system depending on the energy requirements. When $E_{user}$ is less than $KQ_{user}$, FEL mode will be adopted: PGU generates all electricity for users and the auxiliary boiler will supplement the remaining thermal request. On the contrary, if $E_{user}$ is larger than $KQ_{user}$, FTL will be selected: the heat recovery system will supply total thermal request and the lack of electricity will be purchased from the power grid. Except the energy requirements, other criteria are not incorporated into HETS. Essentially, HETS does not use the EC to determine the operating mode.

Substituting related parameters into (2.24), we obtain $K$=0.417. From (2.26), an Equal-Load Interface is determined as $E_{user}$=0.521$Q_h$+0.595$Q_c$. As demonstrated in Figure 2.5(a) with gray shading, the Part-I region of this Interface corresponds to Case 1, $E_{user} \geq 0.417Q_{user}$; the Part-II region corresponds to Case 2, $E_{user} < 0.417Q_{user}$. According to HETS, with the changes of energy requirements, the CCHP system

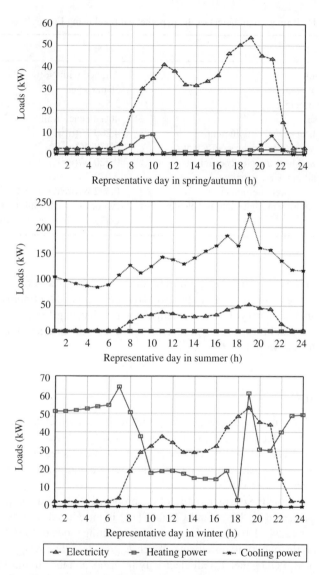

**Figure 2.4** Hourly cooling, heating and power loads of the hypothetical hotel in representative days of spring/autumn, summer and winter

should be operated in FTL mode in the Part-I region and in FEL mode in the Part-II region, respectively.

Furthermore, according to the analysis in Section 2.4, if we can obtain some solutions to make (2.33) and/or (2.36) equal to 0, the Equal-Mode Interfaces will be determined in the Part-I region and/or Part-II region in the operating space. But after some calculations, we find that the value of (2.36) is always less than 0 (all of the

**Table 2.2**  System coefficients

| Symbol | Variable | Value |
|--------|----------|-------|
| $\eta_{pgu}$ | Efficiency of the PGU | 0.25 |
| $\eta_{hrs}$ | Efficiency of the heat recovery system | 0.8 |
| $COP_{ac}$ | COP of the absorption chiller | 0.7 |
| $\eta_h$ | Efficiency of the heating unit | 0.8 |
| $\eta_b$ | Efficiency of the boiler | 0.8 |
| $COP_{ec}$ | COP of the electric chiller | 3.0 |
| $C_{ca}$ | Carbon tax rate (Yuan/kWh) | 0.00002 |
| $\mu_e$ | $CO_2$ emission conversion factor of electricity (g/kWh) | 968 |
| $\mu_f$ | $CO_2$ emission conversion factor of natural gas (g/kWh) | 220 |
| $k_e$ | Site-to-primary electricity conversion factor | 3.336 |
| $k_f$ | Site-to-primary natural gas conversion factor | 1.047 |
| $C_f$ | Natural gas rate (Yuan/kWh) | 0.19 |
| $C_e$ | Electricity rate at 6:00–21:00 (Yuan/kWh) | 0.93 |
| $C_e$ | Electricity rate at 22:00–5:00 (Yuan/kWh) | 0.55 |
| $C_s$ | Electricity sold-back rate (Yuan/kWh) | 0.00 |

related parameters are positive and $1 - \eta_b/[\eta_{hrs}(1 - \eta_{pgu})] = -0.333 < 0$). It implies that there is no Equal-Mode Interface in the Part-II region in the operating space, and the operating mode should always be FEL when $E_{user} < 0.417Q_{user}$.

Referring to the current business electricity prices in Beijing, an average purchased price is selected as $C_e$=0.93 Yuan/kWh and the sold price is set as $C_s$=0 Yuan/kWh. (Currently, the small-scale and medium-scale self-supply energy systems are not allowed to sell electricity to the grid.) By setting (2.33) to be equal to 0, we get $E_{user} = 1.642Q_h - 0.333Q_c$. With this equation, an Equal-Mode Interface is drawn in the Part-I region in Figure 2.5(b). Lines are used to distinguish it from Equal-Load Interface. And then, the Part-I region is divided into two sub-regions: Part I-1 and Part I-2. If the operating point of the CCHP system is in Part I-1, the value of (2.33) will be greater than 0. It means that FTL mode should be adopted as the current operating mode. In comparison, in Part I-2, FEL mode should be adopted. Clearly, the angle of the Equal-Mode Interface and the Equal-Load Interface will be changed with different electricity prices.

To illustrate the feasibility and effectiveness of the proposed optimal switching operation strategy, some simulation tests are presented according to the representative days' electric and thermal requirements, as shown in Figure 2.4. HETS is also used as a comparing strategy in these tests. In addition, the time-of-use electricity price is used: from 6:00 to 21:00, the price is 0.93 Yuan/kWh; from 22:00 to 5:00, the price is 0.55 Yuan/kWh. The changing of operating modes and the amount of energy supply are compared in Figure 2.6.

In Figure 2.6, the first two pairs of figures consistently show that, regardless of the optimal strategy type, the results are consistent for representative days in spring, summer, and autumn.

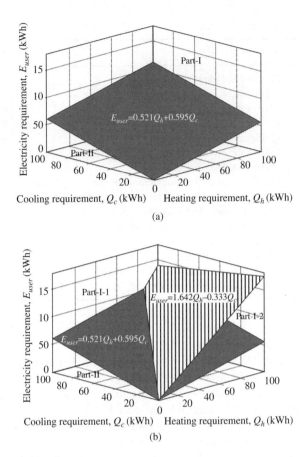

**Figure 2.5**  Space division of operating modes for the hypothetical CCHP system: (a) Equal-Loads Interface; and (b) Equal-Modes Interface

However in winter, because of the impact of EC, the proposed strategy conducts a further division of the operating space. The optimized operating mode and the energy supply are almost completely different from those of HETS from 9:00 to 21:00. These situations are shown in the bottom pair of figures in Figure 2.6.

In particular, the negative values of thermal energy from the auxiliary boiler do not indicate the insufficiency of supplementary energy. On the contrary, it means that the supplementary energy has exceeded the requirement of users. Therefore, in order to further improve the utilization efficiency of the CCHP system, this redundant energy should be stored and reused in appropriate ways. Regarding the thermal energy storage, Piacentino and Cardona [18] and Wang et al. [19] have carried out some valuable research.

To more clearly explain the difference between two optimal operation strategies, we use operating data of each day in the winter heating season (November 15 –March 15) to calculate and compare the performance criteria of the SP system and the CCHP

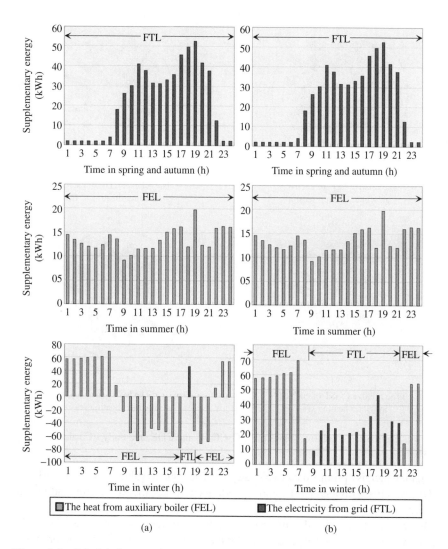

**Figure 2.6** Scheduled status of operating modes and energy supply for representative days' energy requirements of the hypothetical CCHP system: (a) scheduled by the proposed strategy; and (b) scheduled by the HETS

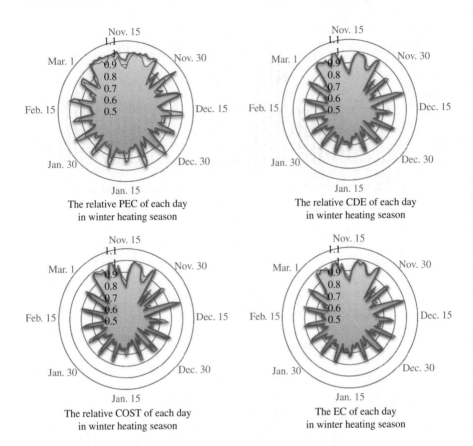

**Figure 2.7** Relative values of PEC, COST, and CDE, and daily EC values

system, which is optimized by two strategies. The relative values of PEC, COST, and CDE, which are divided by the corresponding criteria of the SP system, and EC are shown in Figure 2.7.

In order to distinguish the difference clearly, the criterion curves under the HETS strategy are shaded with light gray, and the comparative curves under the proposed

**Table 2.3** Performance criteria of the whole heating season with different systems and optimal operation strategies

| System | PEC (kWh) | COST (Yuan) | CDE (g) | EC | Stored redundant heat (kWh) |
|---|---|---|---|---|---|
| SP | 368 225.4 | 89 245.35 | 95 320 459 | — | 0 |
| CCHP (HETS) | 302 723.2 | 68 174.11 | 73 897 361 | 0.831339 | 0 |
| CCHP (Proposed) | 316 326.3 | 65 746.13 | 72 496 865 | 0.830792 | 37 703.3 |

strategy are emphasized with bold lines. The overall statistical values of the criteria are listed in Table 2.3.

Evidently, for the evaluated CCHP system, the comprehensive performance with the proposed strategy is better than that with HETS, although the PEC criterion is higher. In addition, more than 37 700 kWh equivalent redundant heat has been produced by applying the proposed strategy. If energy storage devices are installed, the efficiency of the CCHP system will be further improved.

## 2.6 Summary

In this chapter, a novel optimal switching operation strategy for the CCHP system is proposed. Using this strategy, the whole operating space of the CCHP system can be divided into several regions by two or three border surfaces considering not only energy requirements but also the EC. And then the operating point of the CCHP system can be located in a corresponding operating mode region for better overall performance. A CCHP system in a hypothetical hotel building is constructed to evaluate the proposed strategy. Results demonstrated that, for the same CCHP system in the hypothetical hotel, the EC with the proposed strategy is better than that with the comparison strategy, and the new strategy can effectively reflect and balance the influences of consumption, cost, and emissions.

The proposed EC can also flexibly characterize the effects from policies and markets' variations by adjusting the weighting coefficients dynamically. In practice, the adjusting frequency of weighting coefficients and working frequency of the optimal strategy should be in different time scales. For example, if the working frequency is per hour or every several hours, the adjusting frequency should be weekly or monthly. This setting can not only decrease the computational load but also reduce the impact on regular operation of the CCHP system.

Although the efficiency drops of the CCHP equipment at partial load operation are neglected in the design of the proposed optimal operation strategy, some parameters related to the equipment efficiency are still included in the optimal decision functions. Therefore, timely detection and replacement of efficiency parameters will effectively improve the performance of the strategy.

## References

[1] P. J. Mago, N. Fumo, and L. M. Chamra, "Performance analysis of CCHP and CHP systems operating following the thermal and electric load," *International Journal of Energy Research*, vol. 33, no. 9, pp. 852–864, 2009.

[2] C. Sondergren and H. Ravn, "A method to perform probabilistic production simulation involving combined heat and power units," *IEEE Transactions on Power Systems*, vol. 11, no. 2, pp. 1031–1036, 1996.

[3] E. Cardona and A. Piacentino, "A methodology for sizing a trigeneration plant in mediterranean areas," *Applied Thermal Engineering*, vol. 23, no. 13, pp. 1665–1680, 2003.

[4] N. Fumo and L. M. Chamra, "Analysis of combined cooling, heating, and power systems based on source primary energy consumption," *Applied Energy*, vol. 87, no. 6, pp. 2023–2030, 2010.

[5]   G. Chicco and P. Mancarella, "Assessment of the greenhouse gas emissions from congeneration and trigeneration systems. Part I: Models and indicators," *Energy*, vol. 33, no. 3, pp. 410–417, 2008.

[6]   P. Mancarella and G. Chicco, "Assessment of the greenhouse gas emissions from cogeneration and trigeneration systems. Part II: Analysis techniques and application cases," *Energy*, vol. 33, no. 3, pp. 418–430, 2008.

[7]   N. Fumo, P. J. Mago, and L. M. Chamra, "Emission operational strategy for combined cooling, heating, and power systems," *Applied Energy*, vol. 86, no. 11, pp. 2344–2350, 2009.

[8]   J. Wang, C. Zhang, and Y. Jing, "Multi-criteria analysis of combined cooling, heating and power systems in different climate zones in China," *Applied Energy*, vol. 87, no. 4, pp. 1247–1259, 2010.

[9]   E. Cardona, A. Piacentino, and F. Cardona, "Matching economical, energetic, and environmental benefits: An analysis for hybrid CCHP-heat pump systems," *Energy Conversion and Management*, vol. 47, no. 20, pp. 3530–3542, 2006.

[10]  H. Cho, P. J. Mago, R. Luck, and L. M. Chamra, "Evaluation of CCHP systems performance based on operational cost, primary energy concumption, and carbon dioxide emission by utilizing an optimal operation scheme," *Applied Energy*, vol. 86, no. 12, pp. 2540–2549, 2009.

[11]  E. Cardona, A. Piacentino, and F. Cardona, "Energy saving in airports by trigeneration. Part I: Assessing economic and technical potential," *Applied Thermal Engineering*, vol. 26, no. 14–15, pp. 1427–1436, 2006.

[12]  E. Cardona, P. Sannino, A. Piacentino, and F. Cardona, "Energy saving in airports by trigeneration. Part II: Short and long term planning for the Malpensa 2000 CHCP plant," *Applied Thermal Engineering*, vol. 26, no. 14–15, pp. 1437–1447, 2006.

[13]  A. Zafra-Cabeza, M. A. Ridao, I. Alvarado, and E. F. Camacho, "Applying risk management to combined heat and power plants," *IEEE Transactions on Power Systems*, vol. 23, no. 3, pp. 938–945, 2008.

[14]  P. J. Mago and L. M. Chamra, "Analysis and optimization of CCHP systems based on energy, economical, and environmental considerations," *Energy and Buildings*, vol. 41, no. 10, pp. 1099–1106, 2009.

[15]  A. Smith, R. Luck, and P. J. Mago, "Analysis of a combined cooling, heating, and power system model under different operating strategies with input and model data uncertainty," *Energy and Buildings*, vol. 42, no. 11, pp. 2231–2240, 2010.

[16]  G. Chicco and P. Mancarella, "From cogeneration to trigeneration: Profitable alternatives in a competitive market," *IEEE Transactions on Energy Conversion*, vol. 21, no. 1, pp. 265–272, 2006.

[17]  A. Rong and R. Lahdelma, "An efficient linear programming model and optimization algorithm for trigeneration," *Applied Energy*, vol. 82, no. 1, pp. 40–63, 2005.

[18]  A. Piacentino and F. Cardona, "EABOT –energetic analysis as a basis for robust optimization of trigeneration systems by linear programming," *Energy Conversion and Management*, vol. 49, no. 11, pp. 3006–3016, 2008.

[19]  J. Wang, Z. Zhai, Y. Jing, and C. Zhang, "Particle swarm optimization for redundant building cooling heating and power system," *Applied Energy*, vol. 87, no. 12, pp. 3668–3679, 2010.

[20]  D. W. Wu and R. Z. Wang, "Combined cooling, heating and power: A review," *Progress in Energy and Combustion Science*, vol. 32, no. 5-6, pp. 459–495, 2006.

[21]  A. Piacentino and F. Cardona, "On thermoeconomics of energy systems at variable load conditions: Integrated optimization of plant design and operation," *Energy Conversion and Management*, vol. 48, no. 8, pp. 2341–2355, 2007.

[22]  P. Mancarella and G. Chicco, "Global and local emission impact assessment of distributed cogeneration systems with partial-load models," *Applied Energy*, vol. 86, no. 10, pp. 2096–2106, 2009.

[23]  R. Dawes and B. Corrigan, "Linear models in decision making," *Psychological Bulletin*, vol. 81, no. 2, pp. 95–106, 1974.

[24]  US Department of Energy. EnergyPlus. [Online]. Available: https://www.energyplus.net. Accessed February 7, 2017.

# 3

# A Balance-Space-Based Operation Strategy for CCHP Systems

## 3.1 Introduction and Related Work

Performance and efficiencies of CCHP systems mainly depend on system structures, operation strategies, and choices of facility capacity. In [1, 2], the authors proposed a new CCHP system structure different from the conventional one, whose cooling load all lands on the absorption chiller. The new structure adopts a combination of absorption and electric chillers. An electric chiller has a high COP of around 3, which leads to a high cooling efficiency. However, taking the high rates of the electricity into account, an operation strategy needs to be designed to determine the optimal effort of the electric chiller. A parameter called electric cooling to cool load ratio is the one to determine the effort. In the literature [1, 2], this ratio is chosen to be fixed. In contrast, in this chapter, the ratio is optimized according to the variant energy consumption in every hour and the energy rates.

Another problem in designing CCHP systems is the sizing problem, that is, determining the facility capacities. Facilities' capacities in the SP system and chillers, heating unit and boiler in the CCHP system are easy to choose because their sizes solely depend on the corresponding energy load. Some cost-based sizing optimization approaches have been investigated in [3, 4, 5, 6, 7, 8, 9]. However, as a result of the complexity of operation strategies, the PGU capacity is hard to determine. Considering the prices of facilities, external electricity and fuel rates, and facilities' lives, some optimization approaches have been adopted to obtain the optimal PGU capacity, such as the particle swarm optimization [2], the GA optimization [1, 10], and the MINLP algorithm [11, 12]. In this chapter, the enumeration algorithm is adopted to determine the optimal value of PGU capacity.

*Combined Cooling, Heating, and Power Systems: Modeling, Optimization, and Operation*, First Edition.
Yang Shi, Mingxi Liu, and Fang Fang.
© 2017 John Wiley & Sons Ltd. Published 2017 by John Wiley & Sons Ltd.

This chapter is organized in the following way. A description of energy flow of the CCHP system, and the optimal operation strategies for the CCHP system with unlimited and limited PGU capacities are presented in Section 3.2. The EC function used in choosing different strategies constrained by primary energy rates is shown in Section 3.3. The last subsection in Section 3.3 gives the mathematical model of the optimization problem. A case study in Section 3.4 verifies the feasibility of the proposed optimal operation strategies and optimal PGU capacity. Finally, Section 3.5 concludes this chapter.

## 3.2    Optimal Operation Strategy

The system diagram of the CCHP system with hybrid chillers implemented is shown in Figure 3.1. The solid line, dashed line, and dot dashed line represent the thermal energy flow, primary energy flow, and electricity flow, respectively.

The system equations and optimal operation strategies for CCHP systems are elaborated in the following two subsections.

### 3.2.1    CCHP Systems with Unlimited PGU Capacity

In this subsection, the scenario in which the PGU has no capacity limitation will be discussed. The idea of the optimal strategy is to make the electric demand and thermal demand match with each other [13]. The term "match" here indicates that all of the electric and thermal demands of the building are provided by the PGU, namely $F_b = 0$, $E_{grid} = 0$ and electric cooling to cool load ratio $x \in [0, 1]$. In most cases, according to different energy consumption in different seasons and different hours

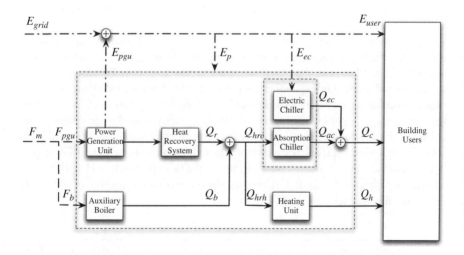

**Figure 3.1**    The CCHP system with hybrid chillers implemented

in one day, the PGU cannot provide the electricity and thermal energy in the exact amount required for the building use. For example, following the FEL strategy, if the thermal load is less than that provided by the PGU, energy waste is inevitable; if the thermal load is larger than the amount PGU provides, additional fuel needs be purchased for the auxiliary boiler. For a CCHP system with a fixed electric cooling to cool load ratio, electric load, cooling load and heating load can only match at a point. However, by introducing a dynamic electric cooling to cool load ratio, the range for the energy loads to match with each other can be extended. By adjusting the electric cooling to cool load ratio, the electric load and thermal load can be accordingly tuned. Since $x$ can only vary in $[0, 1]$, there exists an appropriate range for the $E_{user}$ to vary to match with $Q_c$, $Q_h$, and $E_p$. This range can be obtained by two limiting situations.

First, when $x = 0$, the absorption chiller takes all the cooling load. Thus we have

$$Q_{ac} = Q_c \qquad (3.1)$$

and

$$Q_{hrc} = \frac{Q_{ac}}{COP_{ac}}, \qquad (3.2)$$

where $COP_{ac}$ is the absorption chiller's COP. At the heating unit node, we have

$$Q_{hrh} = \frac{Q_h}{\eta_h}, \qquad (3.3)$$

where $\eta_h$ is the heating efficiency of the heating unit. Then it can be readily derived as

$$Q_r = Q_{hrc} + Q_{hrh}$$
$$= \frac{Q_c}{COP_{ac}} + \frac{Q_h}{\eta_h}. \qquad (3.4)$$

Having the heat recovery system efficiency $\eta_{hrs}$ and the PGU thermal efficiency $\eta_{pgu}$, the fuel consumption $F_{pgu}$ can be calculated as

$$F_{pgu} = \frac{Q_r}{(1 - \eta_{pgu})\eta_{hrs}}, \qquad (3.5)$$

and electricity provided by the PGU is

$$E_{pgu} = \frac{Q_r}{(1 - \eta_{pgu})\eta_{hrs}}\eta_{pgu}. \qquad (3.6)$$

The PGU efficiency varies depending on the PGU load. Usually, low PGU load results in low PGU efficiency. Represented in a second-order polynomial, $\eta_{pgu}$ is

$$\eta_{pgu} = af^2 + bf + c, \qquad (3.7)$$

where $f$ is the fraction of the PGU electric load. If the maximum electric load the PGU can take is $\bar{E}_o^{pgu}$, then we have

$$f = \frac{\min\{E_{user} + E_p + xQ_{ec}/COP_{ec}, \bar{E}_o^{pgu}\}}{\bar{E}_o^{pgu}}. \tag{3.8}$$

To avoid the low system efficiency caused by the low PGU efficiency, a threshold $\gamma$ of the electric load fraction should be set to be an on–off coefficient of the PGU, that is, if $f < \gamma$, the PGU will be shut down to keep a high system efficiency.

If the electric and thermal demands match, $E_{pgu}$ in (3.6) can meet the electric demand. $E_{pgu}$ can be represented as

$$\begin{aligned} E_{pgu} &= E_p + E_{user} \\ &= \frac{Q_c/COP_{ac} + Q_h/\eta_h}{(1 - \eta_{pgu})\eta_{hrs}}\eta_{pgu}. \end{aligned} \tag{3.9}$$

Thus, when $x = 0$, we have

$$E_{user} = \frac{Q_h/\eta_h}{(1 - \eta_{pgu})\eta_{hrs}}\eta_{pgu} + \frac{Q_c/COP_{ac}}{(1 - \eta_{pgu})\eta_{hrs}}\eta_{pgu} - E_p, \tag{3.10}$$

and further let $E_{useru}$ denote the value calculated from (3.10).

In the next step, $x$ is set to be 1, implying that the electric chiller takes all the cooling load and the absorption chiller is left idle. Hence

$$Q_r = \frac{Q_h}{\eta_h}, \tag{3.11}$$

and the fuel consumption $F_{pgu}$ can be readily given by

$$F_{pgu} = \frac{Q_h/\eta_h}{(1 - \eta_{pgu})\eta_{hrs}}. \tag{3.12}$$

At the cooling node, the electric chiller fully provides the cooling demand of the building. Thus $Q_{ec} = Q_c$ and

$$E_{ec} = \frac{Q_c}{COP_{ec}}. \tag{3.13}$$

Since electric and thermal demands match, $E_{pgu}$ provided by $F_{pgu}$ can meet the electric demand including the part required by the electric chiller. Thus we have

$$E_{pgu} = E_p + E_{ec} + E_{user} \tag{3.14a}$$

$$= \frac{Q_h/\eta_h}{(1 - \eta_{pgu})\eta_{hrs}}\eta_{pgu}. \tag{3.14b}$$

Then $E_{user}$ can be represented as

$$E_{user} = \frac{Q_h/\eta_h}{(1 - \eta_{pgu})\eta_{hrs}}\eta_{pgu} - \frac{Q_c}{COP_{ec}} - E_p. \tag{3.15}$$

and further let $E_{userl}$ denote the value calculated from (3.15).

Obviously, there exist three situations to discuss, say, given a set of $Q_c$, $Q_h$, $E_p$, and $E_{user}$, $E_{userl} \leq E_{user} \leq E_{useru}$, $E_{user} > E_{useru}$, and $E_{user} < E_{userl}$. In the rest of the chapter, if no specific statement, $E_p$ is assumed to be a constant. Without the loss of generality, we set $E_p = 0$.

#### 3.2.1.1 $E_{userl} \leq E_{user} \leq E_{useru}$

In this case, no electricity from the grid ($E_{grid} = 0$) and extra fuel need to be purchased. Only $x$ is required to change to make the electric and thermal demands match. This can be regarded as the combination of the FEL and FTL strategies. The space for the point ($Q_c, Q_h, E_{user}$) to vary in this case is shown in Figure 3.2, like point $\alpha$, which is the one between the two planes $E_{useru}$ and $E_{userl}$. The angle between the two planes may vary according to different $\eta_{pgu}$ under different electric loads.

#### 3.2.1.2 $E_{user} > E_{useru}$

This case implies $E_{user} > E_{useru}$ implies that, even though $x$ is adjusted to 0 to reduce the electric load, PGU still cannot fulfill electric and thermal demands. The demand point is above the $E_{useru}$ plane in Figure 3.2, for example the point $\beta$. Thus the FTL strategy is adopted with $x$ set to be 0. From (3.1)–(3.9), we have

$$E_{grid} = E_{user} + E_p - \frac{Q_h/\eta_h}{(1 - \eta_{pgu})\eta_{hrs}}\eta_{pgu} - \frac{Q_c/COP_{ac}}{(1 - \eta_{pgu})\eta_{hrs}}\eta_{pgu}. \tag{3.16}$$

#### 3.2.1.3 $E_{user} < E_{userl}$

This case implies that even though electric chiller runs at full capacity, the electricity provided by the PGU still exceeds the demand. The demand point, like point $\gamma$ in

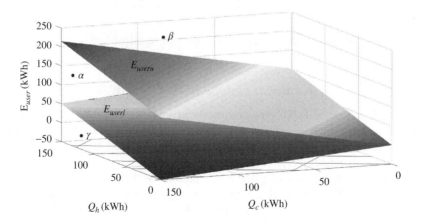

**Figure 3.2**  Space of $Q_c$, $Q_h$, and $E_{user}$

Figure 3.2, is below the plane $E_{userl}$. Because of the much higher COP of the electric chiller, $x$ is set to be 1 to make the electric chiller run at full capacity, then reduce the corresponding $F_{pgu}$; the thermal gap is then compensated for by the auxiliary boiler. The redundant electricity $E_{red}$ is expressed as

$$E_{red} = \frac{Q_h/\eta_h}{(1 - \eta_{pgu})\eta_{hrs}}\eta_{pgu} - \left(\frac{Q_c}{COP_{ec}} + E_p + E_{user}\right), \tag{3.17}$$

and the corresponding redundant fuel $F_{red}$ is

$$F_{red} = \frac{E_{red}}{\eta_{pgu}}, \tag{3.18}$$

which reveals that the thermal gap $Q_{gap}$ is

$$Q_{gap} = F_{red}(1 - \eta_{pgu})\eta_{hrs}. \tag{3.19}$$

This thermal gap will be compensated for by the auxiliary boiler. Fuel needed to feed the auxiliary boiler can be calculated as

$$F_b = \frac{Q_{gap}}{\eta_b}, \tag{3.20}$$

where $\eta_b$ is the efficiency of the auxiliary boiler. From (3.17)–(3.20), we have

$$F_b = \frac{Q_h/\eta_h}{\eta_b} - \left(\frac{Q_c}{COP_{ec}} + E_p + E_{user}\right)\frac{(1 - \eta_{pgu})\eta_{hrs}}{\eta_{pgu}\eta_b}. \tag{3.21}$$

$F_{pgu}$ will be reduced by $F_{red}$, which means

$$\begin{aligned}
F_{pgu} &= \frac{Q_h/\eta_h}{(1 - \eta_{pgu})\eta_{hrs}} - F_{red} \\
&= \frac{Q_c/COP_{ec} + E_p + E_{user}}{\eta_{pgu}}.
\end{aligned} \tag{3.22}$$

In this case, the strategy can be regarded as the FEL strategy.

## 3.2.2 CCHP Systems with Limited PGU Capacity

In this subsection, the PGU being discussed can only provide limited electricity and thermal energy, which means that we cannot simply search for the optimal $x$ to make demands match, but need to take the installed capacity of PGU into further consideration. Given that the PGU runs at full capacity, eight situations are to be discussed. In the following sub-sections, $Q_{pro}$ and $E_{pro}$ represent the thermal energy and electricity provided by PGU, respectively; $Q_{req}$ and $E_{req}$ represent the thermal and electric demands of the building and the electric chiller, respectively. The optimal matching operation strategy for each of the eight cases is stated in the following eight sub-sections.

### 3.2.2.1  $Q_{pro} < Q_{req}$ and $E_{pro} < E_{req}$, $\forall x \in [0,1]$

In this scenario, no matter what $x$ is, the PGU can meet neither electric nor thermal demand. Thus the PGU should run at full capacity. Since the hourly EC function ($EC_{hour}$), which will be discussed in Section 3.3, is a monotonous function of $x$, according to different PEC and CDE, two restricted situations need to be compared, say $x = 0$ and $x = 1$.

When $x = 0$, the absorption chiller runs at full capacity. The PGU will provide the maximum electricity and thermal energy as

$$E_{pgu} = F_{pgum}\eta_{pgu}, \tag{3.23}$$

$$Q_r = F_{pgum}(1 - \eta_{pgu})\eta_{hrs}, \tag{3.24}$$

where $F_{pgum}$ is the capacity of the PGU. Then the boiler will be used to compensate for the thermal gap. Fuel consumed by the auxiliary boiler is

$$F_b = \frac{Q_h/\eta_h + Q_c/COP_{ac} - F_{pgum}(1 - \eta_{pgu})\eta_{hrs}}{\eta_b}. \tag{3.25}$$

The electricity shortage should be purchased from local grid which reveals that

$$E_{grid} = E_{user} + E_p - F_{pgum}\eta_{pgu}. \tag{3.26}$$

When $x = 1$, the electric chiller runs in full capacity. The PGU provides the maximum electricity and thermal energy as shown in (3.23) and (3.24). The thermal gap will be compensated for by the heat provided by the auxiliary boiler. Fuel consumption of the auxiliary boiler can be calculated as

$$F_b = \frac{Q_h/\eta_h - F_{pgum}(1 - \eta_{pgu})\eta_{hrs}}{\eta_b}. \tag{3.27}$$

Electricity consumed by the electric chiller $E_{ec}$ is $Q_c/COP_{ec}$. Then the purchased electricity is

$$E_{grid} = E_{user} + E_p + Q_c/COP_{ec} - F_{pgum}\eta_{pgu}. \tag{3.28}$$

After calculating the $EC_{hour}$ function for the two configurations, the one with the larger result is chosen as the optimal strategy for the corresponding hour.

### 3.2.2.2  $Q_{pro} \geq Q_{req}$ and $E_{pro} < E_{req}$, when $x = 0$

In this case, as long as we raise the value of $x$, $Q_{req}$ keeps descending and $E_{req}$ keeps rising. Thus, $\forall x \in [0, 1]$, we have $Q_{pro} \geq Q_{req}$ and $E_{pro} < E_{req}$. The same as the first scenario, the strategy chosen should be determined by primary energy prices and CDE. This time, we have three options to choose from.

First, let the PGU run at full capacity with $x = 0$. With this configuration, heat provided by the PGU exceeds the thermal demand, but the electricity provided is still not enough. Thus, there is no need for the auxiliary boiler to run, but additional electricity must be purchased from the local grid, implying

$$E_{grid} = E_{user} + E_p - F_{pgum}\eta_{pgu},$$ (3.29)

$$F_{pgu} = F_{pgum}.$$ (3.30)

Secondly, set $x = 0$ and let the PGU exactly cover the thermal load, which means that $F_{pgu} < F_{pgum}$ and $F_b = 0$. $F_{pgu}$ can be calculated from the thermal load by

$$F_{pgu} = \frac{Q_c/COP_{ac} + Q_h/\eta_h}{(1 - \eta_{pgu})\eta_{hrs}}.$$ (3.31)

The electricity gap will be compensated for by the electricity purchased from the local grid with

$$\begin{aligned} E_{grid} &= E_{user} + E_p - E_{pgu} \\ &= E_{user} + E_p - \frac{Q_c/COP_{ac} + Q_h/\eta_h}{(1 - \eta_{pgu})\eta_{hrs}}\eta_{pgu}. \end{aligned}$$ (3.32)

Finally, set $x = 1$ and let the PGU exactly cover the thermal load, which means that $F_{pgu} \ll F_{pgum}$ and $F_b = 0$. $F_{pgu}$ can be calculated from the heating load by

$$F_{pgu} = \frac{Q_h/\eta_h}{(1 - \eta_{pgu})\eta_{hrs}}.$$ (3.33)

The electricity gap will be compensated for by the electricity purchased from the local grid with

$$\begin{aligned} E_{grid} &= E_{user} + E_p + Q_{ec} - E_{pgu} \\ &= E_{user} + E_p + \frac{Q_c}{COP_{ec}} - \frac{Q_h/\eta_h}{(1 - \eta_{pgu})\eta_{hrs}}\eta_{pgu}. \end{aligned}$$ (3.34)

After calculating the value of the $EC_{hour}$ function of these three cases, the largest result will be chosen as the optimal operation strategy for the corresponding hour.

### 3.2.2.3 $Q_{pro} < Q_{req}$ and $E_{pro} \geq E_{req}$, when $x = 1$

In this case, as long as we reduce the value of $x$, $Q_{req}$ keeps rising and $E_{req}$ keeps descending. Thus, $\forall x \in [0, 1]$, we have $Q_{pro} < Q_{req}$ and $E_{pro} \geq E_{req}$. Consequently, if the PGU runs at full capacity, no matter what $x$ is, the electricity provided by the PGU exceeds the electric demand, and the thermal energy provided is still not enough. So the auxiliary boiler needs to be started to compensate for the thermal gap. Here we set $x = 1$ to purchase less fuel, and raise the cooling efficiency by using the electric chiller. Fuel required to be purchased is

$$\begin{aligned} F_b &= \frac{Q_{hrh} - Q_r}{\eta_b} \\ &= \frac{Q_h/\eta_h - F_{pgum}(1 - \eta_{pgu})\eta_{hrs}}{\eta_b}. \end{aligned}$$ (3.35)

Obviously, we have $E_{grid} = 0$ and $F_{pgu} = F_{pgum}$ for this strategy.

### 3.2.2.4 $Q_{pro} \geq Q_{req}$ and $E_{pro} \geq E_{req}$, when $x = 0$

If the condition is satisfied $\forall x \in [0, 1]$, then this scenario means that when the PGU runs at full capacity, both the electricity and thermal energy provided by the PGU exceed the demands. Thus we can simply refer to the strategy choosing method in Section 3.2.1. However, this scenario can not only serve this much stronger condition, say $\forall x \in [0, 1]$, but also the relaxed condition, i.e., only $x = 0$. This is because, when $x = 0$, $Q_{pro} \geq Q_{req}$ and $E_{pro} \geq E_{req}$, and when $x = 1$, $Q_{pro} \geq Q_{req}$ and $E_{pro} < E_{req}$ we can always let $F_{pgu} < F_{pgum}$ by reducing $F_{pgu}$ from $F_{pgum}$ and raising $x$ to an appropriate value. By doing so, $F_{pgum}$ can also be regarded as an unlimited PGU capacity. Then the following procedure will follow similar lines as in Section 3.2.1.

### 3.2.2.5 $Q_{pro} = Q_{req}$ and $E_{pro} = E_{req}$, $\exists x \in [0, 1]$

This scenario can be expanded as when $x = 0$, $Q_{pro} < Q_{req}$ and $E_{pro} \geq E_{req}$, and when $x = 1$, $Q_{pro} \geq Q_{req}$ and $E_{pro} < E_{req}$. Since a particular $x \in [0, 1]$ can make electric and thermal demands match, when the PGU runs at full capacity, it satisfies our optimal matching condition stated before. Thus, we choose the calculated $x$ and $F_{pgu} = F_{pgum}$ as the optimal strategy.

### 3.2.2.6 $Q_{pro} < Q_{req}$ and $E_{pro} < E_{req}$, when $x = 1$, and $Q_{pro} < Q_{req}$ and $E_{pro} \geq E_{req}$, when $x = 0$

In this scenario, two options need to be compared. The first one is setting $x = 1$ and purchasing both the electricity from the local grid and the fuel. The second one is to find an appropriate $x \in [0, 1]$, which makes the system meet the electric demand with only fuel purchased. The optimal decision is made depending on the $EC_{hour}$ function value.

$F_b$ and $E_{grid}$ in the first option can be calculated by (3.27) and (3.28).

With the second option, no additional electricity from the local grid needs to be purchased. The electric cooling to cool load $x$ should be adjusted to

$$x = \frac{(F_{pgum}\eta_{pgu} - E_{user})COP_{ec}}{Q_c}, \tag{3.36}$$

and the purchased fuel is

$$F_b = \frac{[Q_c - (F_{pgum}\eta_{pgu} - E_{user})COP_{ec}]/COP_{ac} + Q_h/\eta_h}{\eta_b} - \frac{F_{pgum}(1 - \eta_{pgu})\eta_{hrs}}{\eta_b}. \tag{3.37}$$

By calculating the $EC_{hour}$ function values of the two options, the one with the larger result will be chosen as the optimal strategy for the corresponding hour.

**3.2.2.7**  $Q_{pro} \geq Q_{req}$ and $E_{pro} < E_{req}$, when $x = 1$, and $Q_{pro} < Q_{req}$
and $E_{pro} < E_{req}$, when $x = 0$

In this scenario, no matter what $x \in [0, 1]$ is, the PGU cannot provide enough electricity for the building and facilities to use. However, thermal energy provided by the PGU can exceed the demand when $x$ is set to be 1. Thus, we adjust $x$ to drive the PGU, when operating at full capacity, provide the exact thermal demand and compensate the electricity gap by purchasing electricity from the local grid. $x$ can be calculated as

$$x = 1 - \frac{[F_{pgum}\eta_{hrs}(1 - \eta_{pgu}) - Q_h/\eta_h]COP_{ac}}{Q_c}, \tag{3.38}$$

and electricity purchased from the local grid is

$$E_{grid} = \frac{xQ_c}{COP_{ec}} + E_{user} - F_{pgum}\eta_{pgu}. \tag{3.39}$$

**3.2.2.8**  $Q_{pro} \geq Q_{req}$ and $E_{pro} \geq E_{req}$, when $x = 1$, and $Q_{pro} < Q_{req}$
and $E_{pro} \geq E_{req}$, when $x = 0$

When the PGU runs at full capacity, electricity provided by the PGU always exceeds the electric demand, but the relationship between the heat provided and the thermal demand varies from less than to more than as $x$ increases from 0 to 1. Check the condition

$$\frac{Q_h}{(1 - \eta_{pgu})\eta_{hrs}\eta_h}\eta_{pgu} \geq \frac{Q_c}{COP_{ec}}. \tag{3.40}$$

If the condition can be satisfied, $x$ is set to be 1 and

$$F_{pgu} = \frac{Q_h}{(1 - \eta_{pgu})\eta_{hrs}\eta_h}. \tag{3.41}$$

If not, we can tell that the situation is the same as the fourth scenario. By the same procedure of calculating $x \in [0, 1]$ and the appropriate $F_{pgu}$, which is smaller than $F_{pgum}$, we have the optimal matching strategy chosen.

## 3.3  EC Function Construction

Similar to Section 2.3, the performance EC also consists of three parts, that is, PEC, cost, and $CO_2$ emissions. Different from Chapter 2, in this chapter, we define and use primary energy savings (PES), hourly total cost savings (HTCS), and carbon dioxide emissions reductions (CDER), which will be given in detail later. In addition to the above, for the PEC of the SP system, we start to consider the parasitic electricity hereinafter. Thus, the PEC of the SP system is calculated as

$$F^{SP} = \frac{E_{user} + E_p^{SP}}{\eta_e^{SP}\eta_{grid}} + \frac{Q_c/COP_{ec}}{\eta_e^{SP}\eta_{grid}} + \frac{Q_h/\eta_h}{\eta_b}, \tag{3.42}$$

where $E_p^{SP}$ is the parasitic electricity of the SP system and $\eta_e^{SP}$ is the generation efficiency of the SP system or local grid.

All of the above three criteria are hourly based. In addition, we assume that the capital recovery factor is 1, and that all equipment has an equally long life.

### 3.3.1 PES

PES is the relative primary energy savings of our CCHP system compared with the SP system. The PES value can be calculated as

$$PES \triangleq 1 - \frac{F^{CCHP}}{F^{SP}}. \tag{3.43}$$

### 3.3.2 HTCS

Normally, in the literature, ATCS is adopted to optimize the PGU capacity. In contrast, in this chapter, in order to obtain the optimal operation strategy for every hour, the HTCS is adopted. The HTC value can be calculated as

$$HTC \triangleq E_{grid}C_e + E_{grid}\mu_e C_{ca} + F_m C_f + F_m \mu_f C_{ca} + \frac{\sum_{k=1}^{l} N_k C_k}{8760 \cdot L}, \tag{3.44}$$

where $N_k$ and $C_k$ are the installed capacity of equipment and the initial capital cost of each piece of equipment, respectively; $l$ is the number of pieces of equipment being used and $L$ is the life of a piece of equipment; $C_e$ and $C_f$ are the unit prices of electricity and the fuel, respectively; $C_{ca}$ is the carbon tax rate, and $\mu_e$ and $\mu_f$ are the carbon conversion factor of the electricity and fuel, respectively. Then the HTCS value can be derived as

$$HTCS \triangleq 1 - \frac{HTC^{CCHP}}{HTC^{SP}}. \tag{3.45}$$

### 3.3.3 CDER

The total CDE of an energy system can be calculated as

$$CDE = E_{grid}\mu_e + F_m\mu_f, \tag{3.46}$$

where $\mu_e$ and $\mu_f$ are the carbon conversion factor of electricity and the fuel, respectively. Then the CDER can be computed as

$$CDER \triangleq 1 - \frac{CDE^{CCHP}}{CDE^{SP}}. \tag{3.47}$$

### 3.3.4 EC Function

The hourly EC is defined as

$$EC_{hour} \triangleq \omega_1 PES + \omega_2 HTCS + \omega_3 CDER, \tag{3.48}$$

where $\omega_1$, $\omega_2$, and $\omega_3$ are the weight of PES, HTCS, and CDER, respectively. The boundary condition is $0 \leq \omega_1, \omega_2, \omega_3 \leq 1$ and $\omega_1 + \omega_2 + \omega_3 = 1$.

The annual EC function value is calculated as

$$EC_{annual} \triangleq \sum_{i=1}^{365} \sum_{j=1}^{24} EC_{hour,ij}, \qquad (3.49)$$

where $EC_{hour,ij}$ is the hourly EC function value of day $i$, hour $j$.

### 3.3.5   Optimal PGU Capacity

The capacity of the PGU in CCHP systems is a key factor. As mentioned in the last subsection, all the strategies are related to the PGU capacity. The PGU capacity should not be too small, as this will not be advantageous; meanwhile, it should not be too large either, since larger installed capacity costs more.

The main contribution in this chapter is to adjust $x$ hourly according to different electric and thermal demands. Thus, no global optimal $x$ exists, every $x$ calculated hourly is the optimal one under the electric and thermal demands in the respective hour and a certain PGU capacity. The goal here is to find an optimal PGU capacity in order to obtain the largest annual $EC_{annual}$ function value, that is,

$$\max_{\underbrace{\phantom{xxxx}}_{F_{pgum}}} EC_{annual}. \qquad (3.50)$$

Here, in this chapter, the enumeration algorithm is adopted, that is searching for every significant PGU capacity value, and then choose the one with the largest annual EC function value.

## 3.4   Case Study

### 3.4.1   Hypothetical Building Configuration

In this section, EnergyPlus [14] is chosen to analyze energy consumption of a hypothetical building in Victoria, BC, Canada. The building model is a hypothetical hotel. We assume that the building, which operates all year round, has four floors with a total construction area of 4500 m². The first floor consists of 300 m² dining halls and 825 m² office rooms; the other floors are guest rooms. Construction parameters of the hypothetical hotel are listed in Table 3.1. Figure 3.3 shows the energy consumption in 1 year of the hypothetical hotel.

### 3.4.2   Simulation Parameters

Given the proposed operation strategy and other facilities' capacities, the main goal of this chapter is to find an optimal PGU capacity. The detailed operation strategy has been discussed in the last section. Table 3.2 shows related coefficients in the CCHP system. The fuel we choose here is the natural gas which has been widely used in North America.

**Table 3.1** Construction parameters of the hypothetical hotel

| Variable | Value |
| --- | --- |
| Orientation | Aligned with North |
| Latitude | 48.469°N |
| Longitude | 123.33°W |
| Location | Victoria, BC, Canada |
| Each floor area | 30 m × 37.5 m |
| Each floor height | 3.3 m |
| Glass area | 40% in each wall |
| Glazing heat transfer coefficient | 4.247 W/(m²K) |
| Exterior wall heat transfer coefficient | 0.442 W/(m²K) |
| Interior wall heat transfer coefficient | 0.718 W/(m²K) |
| Floor heat transfer coefficient | 2.930 W/(m²K) |
| Roof heat transfer coefficient | 0.368 W/(m²K) |
| Electric equipment, lights and people densities | According to the public building energy-saving design standard |

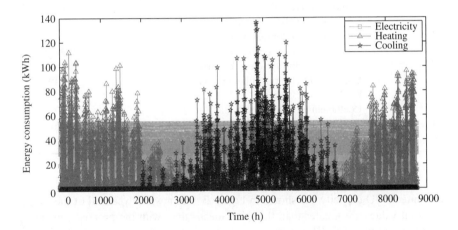

**Figure 3.3** One year energy consumption of a hypothetical hotel in Victoria, BC, Canada

The initial facilities' capacities for optimization, except the PGU capacity, which is the one to be optimized, are chosen according to the simulation data. The initial boilers in the SP system and CCHP system are chosen to be 200 kW and 120 kW, respectively; the initial heating unit is chosen to be 150 kW; the initial absorption and electric chillers are both chosen to be 150 kW.

## 3.4.3 Test Results

The hourly $EC_{hour}$ function value in a whole year for this strategy is calculated to verify the feasibility of the proposed optimal operation strategy of the system with

**Table 3.2**  System coefficients

| Symbol | Variable | Value |
|---|---|---|
| $a$ | First coefficient of $\eta_{pgu}$ | $-0.2$ |
| $b$ | Second coefficient of $\eta_{pgu}$ | 0.4 |
| $c$ | Third coefficient of $\eta_{pgu}$ | 0.1 |
| $\alpha$ | Threshold of the electric load fraction | 25% |
| $\eta_e^{SP}$ | Generation efficiency SP system | 0.35 |
| $\eta_h$ | Efficiency of heating unit | 0.8 |
| $\eta_b$ | Efficiency of boiler | 0.8 |
| $\eta_{hrs}$ | Efficiency of heat recovery system | 0.8 |
| $COP_{ac}$ | Coefficient of performance of absorption chiller | 0.7 |
| $COP_{ec}$ | Coefficient of performance of electric chiller | 3 |
| $\eta_{grid}$ | Transmission efficiency of local grid | 0.92 |
| $\mu_e$ | $CO_2$ emission conversion factor of electricity (g/kWh) | 968 |
| $\mu_f$ | $CO_2$ emission conversion factor of natural gas (g/kWh) | 220 |
| $C_e$ | Electricity rates ($/kWh) | 0.0667 |
| $C_f$ | Natural gas rates ($/kWh) | 0.0516 |
| $C_c$ | Carbon tax rates ($/g) | 0.000003 |
| $C_{pgu}$ | Unit price of PGU ($/kWh) | 1046 |
| $C_b$ | Unit price of boiler ($/kWh) | 46 |
| $C_h$ | Unit price of heating unit ($/kWh) | 30 |
| $C_{ac}$ | Unit price of absorption chiller ($/kWh) | 185 |
| $C_{ec}$ | Unit price of electric chiller ($/kWh) | 149 |
| $L$ | Facilities' lives (year) | 10 |
| $\omega_1$ | Coefficient of PES | 1/6 |
| $\omega_2$ | Coefficient of HTCS | 1/6 |
| $\omega_3$ | Coefficient of CDER | 2/3 |

unlimited PGU capacity. As shown in Figure 3.4, in a whole year, all of the $EC_{hour}$ function values are greater than 0, which means that, with the proposed matching operation strategy, CCHP systems with unlimited PGU capacity perform better than the SP system, thus leading to economic efficiency.

Another case study, aiming to obtain the optimal PGU capacity under the proposed operation strategy, is conducted for the CCHP system with limited PGU capacity. Because of the complexity of the proposed operation strategy, we cannot simply use the classical linear method to obtain the optimal value. In this chapter, adopting the enumeration algorithm, we search for the optimal PGU capacity in [1,500] with the 1-year (8760 h) data. The result of searching is shown in Figure 3.5.

From Figure 3.5, we can readily find out the optimal point:

$$F_{pgumopt} = 96 \text{ kW.} \tag{3.51}$$

Before the PGU capacity reaches 96 kW, the $EC_{annual}$ function value oscillates as capacity rises, which is the result of the complex operation strategy. After reaching the

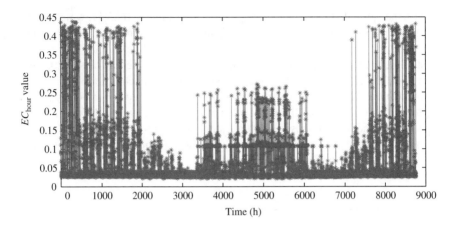

**Figure 3.4**   $EC_{hour}$ function value of CCHP system without capacity limit

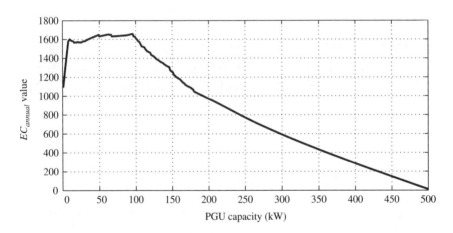

**Figure 3.5**   $EC_{annual}$ function value of different PGU capacities from 1 to 500 kW

optimal capacity, the $EC_{annual}$ function value begins descending as the PGU capacity keeps rising. The reason for descending is that, with the same or even worse performance, the larger the PGU capacity, the higher the cost for the extra PGU installed capacity. Moreover, as the PGU capacity keeps increasing, because of the low electric load, the PGU efficiency can be relative low or even the PGU will be cut down. This can also lead to low system efficiency. In a nutshell, the PGU capacity cannot be too small, which makes the CCHP system lose advantages; neither be so large to reduce the $EC_{annual}$ function value and system efficiency. According to the optimal operation strategy and EC, this optimization approach helps to choose the optimal PGU capacity.

With the optimal PGU capacity, $EC_{annual}$ function value is 1663.57, which implies that the performance of the system can be dramatically improved. The $EC_{hour}$ function value of a whole year is shown in Figure 3.6. It can be observed that, in Figure 3.6, most of the points are between 0.05 and 0.5; few $EC_{hour}$ values are close to 0, which does not affect the performance of the CCHP system too much.

As mentioned before, the electric cooling to cool load ratio varies hourly according to different electric and thermal load. The variation of $x$, with the optimal PGU capacity and under the proposed operation strategy, is shown in Figure 3.7. In winter, spring, and autumn, the value of $x$ switches between 0 and 1; however, in summer, the optimal $x$ varies dramatically between 0 and 1 to obtain the optimal system performance.

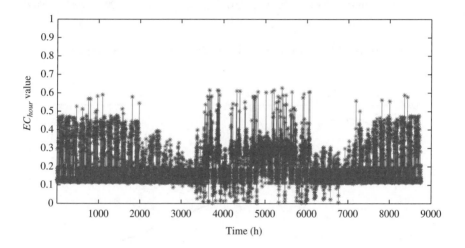

**Figure 3.6**   $EC_{hour}$ function value with 96 kW PGU

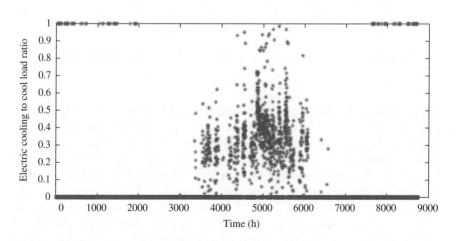

**Figure 3.7**   Variation of electric cooling to cool load ratio in a whole year

**Table 3.3**    EC values of SP and CCHP systems

| System | PEC (kWh) | ATC ($) | CDE (g) |
|--------|-----------|---------|---------|
| SP   | 776 860 | 75 305 | 234 130 000 |
| CCHP | 714 989 | 47 213 | 175 821 242 |
| FEL  | 838 652 | 70 773 | 184 505 056 |
| FTL  | 760 576 | 50 715 | 216 413 613 |

The annual performance criteria of the CCHP system under the proposed optimal operation strategy, FEL strategy, FTL strategy, and SP systems are shown in Table 3.3. From Table 3.3, it can be readily verified that the CCHP system with the proposed operation strategy outperforms the SP system, and CCHP system under FEL and FTL strategies in all performance criteria. In addition, compared with the SP system, the FEL strategy results in less annual cost and CDE, while more PEC. When operating in the FTL strategy, the CCHP system performs better than the SP system; except for the CDE, the PEC and annual cost are less than those of the FEL strategy.

## 3.5   Summary

A new structure of the CCHP system, whose cooling part consists of hybrid chillers, has been constructed in this chapter. The electric cooling to cool load ratio is optimized according to different electric and thermal demands in every respective hour. The energy flow of this CCHP system is investigated as the foundation for the further operation strategy design. By adopting the matching methodology, we first design the operation strategy for the CCHP system with unlimited PGU capacity. To be more practical, an optimal operation strategy, which is based on the relation between full capacity output of the PGU and energy load, is designed for the CCHP system with limited PGU capacity. The enumeration algorithm is adopted to obtain the optimal PGU capacity.

A case study is conducted to show the feasibility of the proposed operation strategies. Test results show that, with the proposed operation strategy and corresponding optimal PGU capacity, the new structured CCHP system performs much better than the conventional SP system in terms of the EC. The ATC of the CCHP system is reduced more than half of that of the SP system.

## References

[1]  J. Wang, Y. Jing, and C. Zhang, "Optimization of capacity and operation for CCHP system by genetic algorithm," *Applied Energy*, vol. 87, no. 4, pp. 1325–1335, 2010.

[2]  J. Wang, Z. Zhai, Y. Jing, and C. Zhang, "Particle swarm optimization for redundant building cooling heating and power system," *Applied Energy*, vol. 87, no. 12, pp. 3668–3679, 2010.

[3]  H. Cho, P. J. Mago, R. Luck, and L. M. Chamra, "Evaluation of CCHP systems performance based on operational cost, primary energy consumption, and carbon dioxide emission by utilizing an optimal operation scheme," *Applied Energy*, vol. 86, no. 12, pp. 2540–2549, 2009.

[4] P. Arcuri, G. Florio, and P. Fragiacomo, "A mixed integer programming model for optimal design of trigeneration in a hospital complex," *Energy*, vol. 32, no. 8, pp. 1430–1447, 2010.

[5] H. Ren, W. Gao, and Y. Ruan, "Optimal sizing for residential CHP system," *Applied Thermal Engineering*, vol. 28, no. 5–6, pp. 514–523, 2008.

[6] B. Zhang and W. Long, "An optimal sizing method for cogeneration plants," *Energy and Buildings*, vol. 38, no. 3, pp. 189–195, 2006.

[7] D. Ziher and A. Poredos, "Economics of a trigeneration system in a hospital," *Applied Thermal Engineering*, vol. 26, no. 7, pp. 680–687, 2006.

[8] G. Chicco and P. Mancarella, "Matrix modelling of small-scale trigeneration systems and application to operational optimization," *Energy*, vol. 34, no. 3, pp. 261–273, 2009.

[9] A. Rong and R. Lahdelma, "An efficient linear programming model and optimization algorithm for trigeneration," *Applied Energy*, vol. 82, no. 1, pp. 40–63–, 2005.

[10] J. Wang, Z. Zhai, Y. Jing, X. Zhang, and C. Zhang, "Sensitivity analysis of optimal model on building cooling heating and power system," *Applied Energy*, vol. 88, no. 12, pp. 5143–5152, 2011.

[11] T. Savola, T.-M. Tveit, and C.-J. Fogelholm, "A MINLP model including the pressure levels and multiperiods for CHP process optimisation," *Applied Thermal Engineering*, vol. 27, no. 11–12, pp. 1857–1867, 2007.

[12] T. Savola and C.-J. Fogelholm, "MINLP optimisation model for increased power production in small-scale CHP plants," *Applied Thermal Engineering*, vol. 27, no. 1, pp. 89–99, 2007.

[13] P. J. Mago, L. M. Chamra, and J. Ramsay, "Micro-combined cooling, heating and power systems hybrid electric-thermal load following operation," *Applied Thermal Engineering*, vol. 30, no. 8–9, pp. 800–806, 2010.

[14] US Department of Energy. EnergyPlus. [Online]. Available: https://www.energyplus.net. Accessed February 7, 2017.

# 4

# Energy Hub Modeling and Optimization-Based Operation Strategy for CCHP Systems

## 4.1 Introduction and Related Work

The CCHP system is the connection between the energy input, that is, the electricity and the fuel, and the building users' demand. From such a mapping perspective, a CCHP system can be viewed as an energy hub with multiple energy vectors at the input and output terminals [1]. The energy hub represents an interface between different energy infrastructures and/or loads. The coupling of different energy carriers is established by the conversions among them. For instance, the PGU in the CCHP system can generate electricity and thermal energy simultaneously by combusting fuel. The electricity and thermal energy provided by the PGU will affect the purchasing of electricity from the local grid and additional fuel for the auxiliary boiler [2, 3]. By using the input–output mapping [4], the CCHP system can be modeled by a connection matrix. In [5], the authors model the system by introducing the concepts of dispatch factors and coupling matrix, and optimize the power flow and operation strategy using the KKT conditions. In [6], the authors model the system by using the concept of junctions, bifurcations, and the backtracking. However, the modeling process of the latter is not practically implementable. In this chapter, a more comprehensive and intuitive approach, that is, the matrix modeling, is proposed to describe the CCHP system. Having the matrix modeled system, the optimal operation strategy can be obtained by solving a non-convex optimization problem. SQP is adopted to solve the problem for a sold-back-disabled system; a novel algorithm is proposed to convexify and solve the problem for a sold-back-enabled system. The result of the optimization problem is the optimal power flow and operation strategy for the CCHP system.

*Combined Cooling, Heating, and Power Systems: Modeling, Optimization, and Operation*, First Edition.
Yang Shi, Mingxi Liu, and Fang Fang.
© 2017 John Wiley & Sons Ltd. Published 2017 by John Wiley & Sons Ltd.

Another equally important problem involved in the CCHP system design is the facility sizing problem, that is, determining the capacities of facilities; here, we only focus on the PGU capacity. Some cost-based sizing problems have been investigated in [7–12], to name a few. On account of the complexity of the operation strategy of the power flow process, the capacity of the PGU is hard to determine. Considering the the price of the PGU, PEC, ATC, and the GHG emissions, several optimization approaches are adopted to obtain the optimal PGU capacity in the literature, such as the particle swarm optimization [13], the GA optimization [14], and the MINLP algorithm [15, 16]. In this chapter, in order to reduce the time and space complexity of the optimization process, we adopt the enumeration algorithm to obtain the optimal PGU capacity, which can minimize the annually objective function.

This chapter is organized in the following way. The matrix modeling approach for the CCHP system is described in Section 4.2, which includes the components' efficiency matrices modeling, dispatch factors definitions and system conversion matrix modeling. Section 4.3 presents the definition of the optimization problem and the constraints of this problem. The case study in Section 4.4 shows the effectiveness and economic efficiency of the proposed optimal power flow and operation strategy. Finally, we conclude this chapter in Section 4.5. The appendix presents the proposed optimization algorithm of convexifying and solving the non-convex problem.

## 4.2 System Matrix Modeling

In this section, a comprehensive and intuitive matrix modeling of the CCHP system will be introduced. The configuration of the CCHP system with hybrid cooling system is shown in Figure 3.1.

### 4.2.1 Efficiency Matrices of System Components

The efficiency matrices, also known as coupling matrices [5], are the description of the energy conversion of the system components. In this chapter, we define the input and output of the $\ell$th component as $\mathcal{V}_i^\ell = [F_i^\ell \ E_i^\ell \ Q_{ci}^\ell \ Q_{hi}^\ell]^\mathsf{T}$ and $\mathcal{V}_o^\ell = [F_o^\ell \ E_o^\ell \ Q_{co}^\ell \ Q_{ho}^\ell]^\mathsf{T}$, respectively. In the rest of this chapter, if there are no specific notes, the elements of the input and output vectors have the same order as $\mathcal{V}_i^\ell$ and $\mathcal{V}_o^\ell$. In the above definition, $F, E, Q_c$ and $Q_h$ denote the fuel, electricity, cooling energy, and heating energy, respectively. Then the input–output relation of the $\ell$th component can be represented as

$$\mathcal{V}_o^\ell = \mathcal{H}^\ell \mathcal{V}_i^\ell, \tag{4.1}$$

where $\mathcal{H}^\ell$ is the efficiency matrix of the $\ell$th component.

The input–output relation of the PGU, whose electric efficiency is $\eta_{pgu}$, can be represented by the following matrix form

$$
\begin{bmatrix} 0 \\ E_o^{PGU} \\ 0 \\ Q_{ho}^{PGU} \end{bmatrix} = \begin{bmatrix} 0 & 0 & 0 & 0 \\ \eta_{pgu} & 0 & 0 & 0 \\ 0 & 0 & 0 & 0 \\ 1 - \eta_{pgu} & 0 & 0 & 0 \end{bmatrix} \begin{bmatrix} F_i^{PGU} \\ 0 \\ 0 \\ 0 \end{bmatrix}
$$

$$
= \mathcal{H}^{PGU} V_i^{PGU}.
$$

(4.2)

Note that, in this chapter, $\eta_{pgu}$ is assumed to be a constant, that is, the PGU efficiency will not vary according to the partial load, for the sake of a better interpretation of the proposed method.

Following the same procedure, the efficiency matrices of the auxiliary boiler, heat recovery system, electric chiller, heating unit, and absorption chiller can be obtained as

$$
\mathcal{H}^b = \begin{bmatrix} 0 & 0 & 0 & 0 \\ 0 & 0 & 0 & 0 \\ 0 & 0 & 0 & 0 \\ \eta_b & 0 & 0 & 0 \end{bmatrix}, \mathcal{H}^{hrs} = \begin{bmatrix} 0 & 0 & 0 & 0 \\ 0 & 0 & 0 & 0 \\ 0 & 0 & 0 & 0 \\ 0 & 0 & 0 & \eta_{hrs} \end{bmatrix},
$$

$$
\mathcal{H}^{ec} = \begin{bmatrix} 0 & 0 & 0 & 0 \\ 0 & 0 & 0 & 0 \\ 0 & COP_{ec} & 0 & 0 \\ 0 & 0 & 0 & 0 \end{bmatrix}, \mathcal{H}^h = \begin{bmatrix} 0 & 0 & 0 & 0 \\ 0 & 0 & 0 & 0 \\ 0 & 0 & 0 & 0 \\ 0 & 0 & 0 & \eta_h \end{bmatrix}.
$$

$$
\mathcal{H}^{ac} = \begin{bmatrix} 0 & 0 & 0 & 0 \\ 0 & 0 & 0 & 0 \\ 0 & 0 & 0 & COP_{ac} \\ 0 & 0 & 0 & 0 \end{bmatrix},
$$

respectively.

## 4.2.2  Dispatch Matrices

The efficiency matrices characterize the performance of the energy conversion inside the components, whereas the dispatch matrices represent the power flow between them. In addition, the dispatch factors only exist at the bifurcations of the system. For example, in Figure 3.1, the fuel supply $F_m$ is separated into two parts: one for the PGU and the other one for the auxiliary boiler. Let $\alpha_{pgu}$ and $\alpha_b$ denote the dispatch factors for the PGU and auxiliary boiler, respectively. Then we have

$$F_{pgu} = \alpha_{pgu} F_m, \tag{4.3}$$

$$F_b = \alpha_b F_m, \tag{4.4}$$

subject to

$$\alpha_{pgu} + \alpha_b = 1. \tag{4.5}$$

For the sake of optimization based operation strategy design, the input vectors of the components can be modified to be the functions of the system input. At the PGU side, we have

$$\mathcal{V}_i^{PGU} = \begin{bmatrix} \alpha_{pgu} & 0 & 0 & 0 \\ 0 & 0 & 0 & 0 \\ 0 & 0 & 0 & 0 \\ 0 & 0 & 0 & 0 \end{bmatrix} \begin{bmatrix} F_m \\ E_{grid} \\ 0 \\ 0 \end{bmatrix} \tag{4.6}$$

$$= \Gamma_{pgu} \mathcal{V}_i,$$

where $\Gamma_{pgu}$ is the dispatch matrix for the PGU.

Next, the dispatch matrices for the auxiliary boiler, electric chiller, heat recovery system, absorption chiller, and heating unit can be generated as

$$\Gamma_b = \begin{bmatrix} \alpha_b & 0 & 0 & 0 \\ 0 & 0 & 0 & 0 \\ 0 & 0 & 0 & 0 \\ 0 & 0 & 0 & 0 \end{bmatrix}, \Gamma_{ec} = \begin{bmatrix} 0 & 0 & 0 & 0 \\ 0 & \alpha_{ec} & 0 & 0 \\ 0 & 0 & 0 & 0 \\ 0 & 0 & 0 & 0 \end{bmatrix},$$

$$\Gamma_{hrs} = \begin{bmatrix} 0 & 0 & 0 & 0 \\ 0 & 0 & 0 & 0 \\ 0 & 0 & 0 & 0 \\ \alpha_{pgu}(1 - \eta_{pgu}) & 0 & 0 & 0 \end{bmatrix},$$

$$\Gamma_{ac} = \begin{bmatrix} 0 & 0 & 0 & 0 \\ 0 & 0 & 0 & 0 \\ 0 & 0 & 0 & 0 \\ [\alpha_{pgu}(1 - \eta_{pgu})\eta_{hrs} + (1 - \alpha_{pgu})\eta_b]\alpha_{ac} & 0 & 0 & 0 \end{bmatrix}$$

$$\Gamma_h = \begin{bmatrix} 0 & 0 & 0 & 0 \\ 0 & 0 & 0 & 0 \\ 0 & 0 & 0 & 0 \\ [\alpha_{pgu}(1 - \eta_{pgu})\eta_{hrs} + (1 - \alpha_{pgu})\eta_b]\alpha_h & 0 & 0 & 0 \end{bmatrix}$$

Similarly, in addition to (4.5), we have

$$\alpha_{user} + \alpha_{ec} = 1,$$
$$\alpha_{ac} + \alpha_h = 1. \tag{4.7}$$

### 4.2.3  Conversion Matrix of the CCHP System

The conversion matrix of the whole CCHP system describes the efficiencies of the components and the whole procedure of the power flow. The operation strategy of the system can also be included inherently in the conversion matrix. From Figure 3.1, we define the system input to be

$$\mathcal{V}_i = \begin{bmatrix} F_m & E_{grid} & Q_{ci} & Q_{hi} \end{bmatrix}^{\mathsf{T}}$$
$$= \begin{bmatrix} F_m & E_{grid} & 0 & 0 \end{bmatrix}^{\mathsf{T}}. \tag{4.8}$$

The second equality in (4.8) holds since neither cooling energy nor heating energy is the input to the whole system. Even though no $Q_{ci}$ and $Q_{hi}$ exist in the system input, we keep their positions to make all the input vectors, including components and the system, in a unified form. The output of the system is defined to be

$$\mathcal{V}_o = \begin{bmatrix} F_o & E_{user} & Q_c & Q_h \end{bmatrix}^{\mathsf{T}}$$
$$= \begin{bmatrix} 0 & E_{user} & Q_c & Q_h \end{bmatrix}^{\mathsf{T}}. \tag{4.9}$$

The second equality in (4.9) holds since no fuel is output from the system. Then the conversion matrix $\mathcal{H}$ of the CCHP system can be defined as

$$\mathcal{V}_o = \mathcal{H}\mathcal{V}_i. \tag{4.10}$$

Without the loss of generality, we assume that the parasitic electricity $E_p = 0$. For the output element $E_{user}$, we have

$$E_{user} = (E_{pgu} + E_{grid})\alpha_{user}$$
$$= \alpha_{user}\alpha_{pgu}\eta_{pgu}F_m + \alpha_{user}E_{grid}. \tag{4.11}$$

As for the cooling part, though electric chillers possess high COP, the unit price of electricity is higher than natural gas. Thus, we believe that it is more efficient to use hybrid chillers. Optimization should be conducted to obtain the optimal electric cooling to cool load ratio, which is inherently included in the modeling procedure. From Figure 3.1, for the cooling part, we have

$$Q_c = Q_{ec} + Q_{ac}$$
$$= \alpha_{pgu}\eta_{pgu}\alpha_{ec}COP_{ec}F_m + \alpha_{ec}COP_{ec}E_{grid} \tag{4.12}$$
$$+ [\alpha_{pgu}(1 - \eta_{pgu})\eta_{hrs} + \alpha_b\eta_b]\alpha_{ac}COP_{ac}F_m.$$

Note that, the electric cooling to cool load ratio is represented by $Q_{ec}/Q_c$.
The last output of the CCHP system is the heating demand which is all provided
by the heating unit as

$$Q_h = (Q_b + Q_r)\alpha_h\eta_h$$
$$= [\alpha_{pgu}(1 - \eta_{pgu})\eta_{hrs} + \alpha_b\eta_b]\alpha_h\eta_h F_m. \tag{4.13}$$

From (4.11) to (4.13), we can readily obtain the conversion matrix as

$$\mathcal{H} = \begin{bmatrix} 0 & 0 & 0 & 0 \\ \alpha_{user}\alpha_{pgu}\eta_{pgu} & \alpha_{user} & 0 & 0 \\ \alpha_{pgu}\eta_{pgu}\alpha_{ec}COP_{ec} + [\alpha_{pgu}(1 - \eta_{pgu})\eta_{hrs} + \alpha_b\eta_b]\alpha_{ac}COP_{ac} & \alpha_{ec}COP_{ec} & 0 & 0 \\ [\alpha_{pgu}(1 - \eta_{pgu})\eta_{hrs} + \alpha_b\eta_b]\alpha_h\eta_h & 0 & 0 & 0 \end{bmatrix}. \tag{4.14}$$

The main objective of this chapter is to determine the dispatch factors and the
system input to minimize the EC objective function.

## 4.3   Optimal Control Design

In this section, we formulate the system design as an optimization problem with
appropriate EC and the corresponding non-linear equality and inequality constraints.

### 4.3.1   Decision Variables

The optimization of the overall CCHP system needs to fulfill the following three
aspects: (1) optimization of the dispatch factors; (2) optimization of the input energy;
and (3) optimization of the PGU capacity. The first part of the optimization is to
coordinate the power flow of the system so as to minimize the objective function.
Optimizing the energy consumption means purchasing a reasonable amount of electricity and fuel to run the CCHP system, in order to meet the building demand and
minimize the objective function. Capacity of the PGU in CCHP systems is also a
key factor. The dispatch factors and energy consumption that are to be optimized are
based on the PGU capacity. PGU capacity should not be too small, which will not be
advantageous; meanwhile, it should not be too large either, as larger installed capacity
costs more.

Since the dispatch factors $\alpha_{pgu}, \alpha_b, \alpha_{user}, \alpha_{ec}, \alpha_{ac}$, and $\alpha_h$ are dependent variables,
the dispatch factors can be reduced to be $\alpha_{pgu}, \alpha_{user}$, and $\alpha_{ac}$, which are independent
of each other. Thus we define the dispatch factor vector to be

$$\boldsymbol{\alpha} = [\alpha_{pgu}\ \alpha_{user}\ \alpha_{ac}]^{\mathsf{T}}. \tag{4.15}$$

In addition, the input energy vector is defined to be

$$\boldsymbol{\beta} = [F_m\ E_{grid}]^{\mathsf{T}}. \tag{4.16}$$

Both $\alpha$ and $\beta$ need to be optimized, so we augment $\alpha$ and $\beta$ to be the optimizer

$$\mathcal{X} = [\alpha^{\mathsf{T}} \ \beta^{\mathsf{T}}]^{\mathsf{T}}. \tag{4.17}$$

### 4.3.2   Objective Function

In this chapter, the EC function is constructed in a similar way as in Chapter 3. However, for the sake of solving optimization problems, we drop the usage of "savings", instead directly using the quotient of the criteria from the CCHP system and criteria from the SP system. Consequently, the objective function can be rewritten as a linear function of the optimizer $\mathcal{X}$ as

$$EC_{hour}(\mathcal{X}) = \omega_1 \frac{P\mathcal{X}}{FSP} + \omega_2 \frac{C\mathcal{X} + \mathcal{L}}{HTC^{SP}} + \omega_3 \frac{D\mathcal{X}}{CDE^{SP}}, \tag{4.18}$$

where

$$P = \begin{bmatrix} 0 & 0 & 0 & 1 & 1/(\eta_e^{SP}\eta_{grid}) \end{bmatrix},$$

$$C = \begin{bmatrix} 0 & 0 & 0 & C_f + \mu_f C_c & C_e + \mu_e C_e \end{bmatrix},$$

$$D = \begin{bmatrix} 0 & 0 & 0 & \mu_f & \mu_e \end{bmatrix},$$

$$\mathcal{L} = \frac{\sum_{k=1}^{l} N_k C_k}{8760 \cdot L}.$$

Then if given a specific PGU capacity, the hourly optimization problem becomes

$$\underbrace{\min}_{\mathcal{X}} \left\{ \omega_1 \frac{P\mathcal{X}}{FSP} + \omega_2 \frac{C\mathcal{X} + \mathcal{L}}{HTC^{SP}} + \omega_3 \frac{D\mathcal{X}}{CDE^{SP}} \right\}, \tag{4.19}$$

and is subject to non-linear equality and inequality constraints. This objective function is a comprehensive one, which integrates three aspects, that is, PEC, hourly total cost (HTC), and CDE, into it. Three weights, $\omega_1$, $\omega_2$, and $\omega_3$, are selected according to different requirements. For example, if more emphasis needs to be put on the CDER, then $\omega_3$ should be raised and $\omega_1$ and $\omega_2$ should be accordingly decreased.

### 4.3.3   Non-linear Equality Constraint

The non-linear equality constraint represents the supply–demand balance, including the fuel, electricity, cooling energy, and heating energy. The only equality constraint for the optimization problem is

$$\mathcal{H}\mathcal{V}_i - \mathcal{V}_o = 0. \tag{4.20}$$

There is no need to restrict the supply and demand relation of each component for this type of constraint is already included in the conversion matrix $\mathcal{H}$.

Since the elements of the conversion matrix $\mathcal{H}$ are represented by the elements of the matrix $\mathfrak{X}$, $\mathcal{H}$ can be rewritten to be a function of $\mathfrak{X}$. Then we have

$$
\begin{aligned}
\mathcal{H} = {} & (h_{11}W_{22} + h_{311}W_{33} + h_{313}W_{32} - h_{411}W_{43})\mathfrak{X}\mathfrak{X}^{\mathsf{T}}U_{11} \\
& + (h_{312}W_{33} + h_{314}W_{31} + h_{411}W_{41} - h_{412}W_{43})\mathfrak{X}Q_{11} \\
& + (h_{221}W_{22} + h_{321}W_{32})\mathfrak{X}Q_{12} \\
& + (CON_{COP_{ec}} + COP_{\eta_h\eta_b}),
\end{aligned}
\tag{4.21}
$$

where

$$
W_{22} = \begin{bmatrix} 0 & 0 & 0 & 0 & 0 \\ 0 & 1 & 0 & 0 & 0 \\ 0 & 0 & 0 & 0 & 0 \\ 0 & 0 & 0 & 0 & 0 \end{bmatrix}, W_{31} = \begin{bmatrix} 0 & 0 & 0 & 0 & 0 \\ 0 & 0 & 0 & 0 & 0 \\ 1 & 0 & 0 & 0 & 0 \\ 0 & 0 & 0 & 0 & 0 \end{bmatrix},
$$

$$
W_{32} = \begin{bmatrix} 0 & 0 & 0 & 0 & 0 \\ 0 & 0 & 0 & 0 & 0 \\ 0 & 1 & 0 & 0 & 0 \\ 0 & 0 & 0 & 0 & 0 \end{bmatrix}, W_{33} = \begin{bmatrix} 0 & 0 & 0 & 0 & 0 \\ 0 & 0 & 0 & 0 & 0 \\ 0 & 0 & 1 & 0 & 0 \\ 0 & 0 & 0 & 0 & 0 \end{bmatrix},
$$

$$
W_{41} = \begin{bmatrix} 0 & 0 & 0 & 0 & 0 \\ 0 & 0 & 0 & 0 & 0 \\ 0 & 0 & 0 & 0 & 0 \\ 1 & 0 & 0 & 0 & 0 \end{bmatrix}, W_{43} = \begin{bmatrix} 0 & 0 & 0 & 0 & 0 \\ 0 & 0 & 0 & 0 & 0 \\ 0 & 0 & 0 & 0 & 0 \\ 0 & 0 & 1 & 0 & 0 \end{bmatrix},
$$

$$
Q_{11} = \begin{bmatrix} 1 & 0 & 0 & 0 \end{bmatrix}, \qquad Q_{12} = \begin{bmatrix} 0 & 1 & 0 & 0 \end{bmatrix},
$$

$$
U_{11} = \begin{bmatrix} 1 & 0 & 0 & 0 \\ 0 & 0 & 0 & 0 \\ 0 & 0 & 0 & 0 \\ 0 & 0 & 0 & 0 \\ 0 & 0 & 0 & 0 \end{bmatrix}, CON_{\eta_h\eta_b} = \begin{bmatrix} 0 & 0 & 0 & 0 \\ 0 & 0 & 0 & 0 \\ 0 & 0 & 0 & 0 \\ \eta_h\eta_b & 0 & 0 & 0 \end{bmatrix},
$$

$$
CON_{COP_{ec}} = \begin{bmatrix} 0 & 0 & 0 & 0 \\ 0 & 0 & 0 & 0 \\ 0 & COP_{ec} & 0 & 0 \\ 0 & 0 & 0 & 0 \end{bmatrix},
$$

and

$$
\begin{aligned}
& h_{211} = \eta_{pgu}, && h_{221} = 1, \\
& h_{312} = \eta_b COP_{ac}, && h_{313} = -\eta_{pgu}COP_{ec}, \\
& h_{321} = -COP_{ec}, && h_{314} = \eta_{pgu}COP_{ec}, \\
& h_{311} = (\eta_{hrs} - \eta_{pgu}\eta_{hrs} - \eta_b)COP_{ac}, \\
& h_{411} = \eta_h(\eta_{hrs} - \eta_{pgu}\eta_{hrs} - \eta_b), \\
& h_{412} = \eta_h\eta_b.
\end{aligned}
$$

In addition, the system input $\mathcal{V}_i$ should also be represented by $\mathcal{X}$ as

$$\mathcal{V}_i = P\mathcal{X}, \tag{4.22}$$

where

$$P = \begin{bmatrix} 0 & 0 & 0 & 1 & 0 \\ 0 & 0 & 0 & 0 & 1 \\ 0 & 0 & 0 & 0 & 0 \\ 0 & 0 & 0 & 0 & 0 \end{bmatrix}.$$

Then the non-linear equality constraint becomes

$$\begin{aligned}
&(h_{211}W_{22} + h_{311}W_{33} + h_{313}W_{32} - h_{411}W_{43})\mathcal{X}\mathcal{X}^T U_{11}P\mathcal{X} \\
&+ (h_{312}W_{33} + h_{314}W_{31} + h_{411}W_{41} - h_{412}W_{43})\mathcal{X}Q_{11}P\mathcal{X} \\
&+ (h_{221}W_{22} + h_{321}W_{32})\mathcal{X}Q_{12}P\mathcal{X} \\
&+ (CON_{COP_{ec}} + COP_{\eta_h \eta_b})P\mathcal{X} - \mathcal{V}_o = 0.
\end{aligned} \tag{4.23}$$

### 4.3.4  Non-linear Inequality Constraints

The inequality constraints mainly focus on the characters of parameters, capacities of components and lower bound of the components' output. The dispatch factors, say the elements in $\alpha$, should be no less than zero and no more than one. The reason is that if one power flow is divided into several parts at a bifurcation, each part only has a certain percent of the total power flow; and the sum of all the divided power flow should be identical to the original power flow. For the energy consumption, since we assume that no energy can be sold back, the $F_m$ and $E_{grid}$ should be no less than zero. These generate

$$-\alpha \leq 0, \tag{4.24a}$$

$$\alpha - 1 \leq 0, \tag{4.24b}$$

$$-\beta \leq 0. \tag{4.24c}$$

Inequalities in (4.24) are linear inequality constraints, however, they can be dealt with as the special case in non-linear inequality constraints, or the lower and upper bounds for the optimizer.

The capacities of components constrain the upper bound of the components' output. Since, except for the PGU capacity that is to be optimized, the capacities of other components have specific limits, here we only address the upper bound of the PGU. The capacity of PGU is denoted by $F_{pgum}$, thus the upper bound of the PGU output is represented as

$$\overline{\mathcal{V}}_o^{pgu} = \begin{bmatrix} 0 \\ \overline{E}_o^{pgu} \\ 0 \\ \overline{Q}_{ho}^{pgu} \end{bmatrix} = \begin{bmatrix} 0 \\ F_{pgum}\eta_{pgu} \\ 0 \\ F_{pgum}(1 - \eta_{pgu}) \end{bmatrix}. \tag{4.25}$$

The output upper bounds for other components are $\overline{\mathcal{V}}_o^b$, $\overline{\mathcal{V}}_o^{hrs}$, $\overline{\mathcal{V}}_o^{ec}$, $\overline{\mathcal{V}}_o^{ac}$, and $\overline{\mathcal{V}}_o^h$.

Some components have thresholds, which imply that if the expected output of one component is lower than its thresholds, this component should be cut off. The lower bound for the PGU, auxiliary boiler, heating recovery system, electric chiller, absorption chiller, and heating unit are $\underline{\mathcal{V}}_o^{pgu}$, $\underline{\mathcal{V}}_o^b$, $\underline{\mathcal{V}}_o^{hrs}$, $\underline{\mathcal{V}}_o^{ec}$, $\underline{\mathcal{V}}_o^{ac}$, and $\underline{\mathcal{V}}_o^h$, respectively. Without losing generality, we can set those lower bounds to be zero.

With the components' output upper and lower bounds constraints, we can readily have the inequality constraints as

$$\mathcal{H}^\ell \mathcal{V}_i^\ell - \overline{\mathcal{V}}_o^\ell \leq 0, \tag{4.26a}$$

$$\underline{\mathcal{V}}_o^\ell - \mathcal{H}^\ell \mathcal{V}_i^\ell \leq 0. \tag{4.26b}$$

Then from (4.6), (4.7), and (4.22), (4.26) becomes

$$\mathcal{H}^\ell \Gamma_\ell P \boldsymbol{\mathcal{X}} - \overline{\mathcal{V}}^\ell \leq 0, \tag{4.27a}$$

$$\underline{\mathcal{V}}^\ell - \mathcal{H}^\ell \Gamma_\ell P \boldsymbol{\mathcal{X}} \leq 0. \tag{4.27b}$$

By following the similar procedure of deriving (4.21), $\Gamma_\ell$ can also be represented by the function of $\boldsymbol{\mathcal{X}}$ as

$$\Gamma_{pgu} = W_{11} \boldsymbol{\mathcal{X}} Q_{11}, \qquad\qquad \Gamma_b = I_{11} - W_{11} \boldsymbol{\mathcal{X}} Q_{11},$$

$$\Gamma_{hrs} = \gamma_1 W_{41} \boldsymbol{\mathcal{X}} Q_{11}, \qquad\qquad \Gamma_{ec} = I_{22} - W_{22} \boldsymbol{\mathcal{X}} Q_{12},$$

$$\Gamma_{ac} = \gamma_2 W_{43} \boldsymbol{\mathcal{X}} \boldsymbol{\mathcal{X}}^\mathsf{T} U_{11} + \gamma_3 W_{43} \boldsymbol{\mathcal{X}} Q_{11},$$

$$\Gamma_h = -\gamma_2 W_{43} \boldsymbol{\mathcal{X}} \boldsymbol{\mathcal{X}}^\mathsf{T} U_{11} + (\gamma_2 W_{41} - \gamma_3 W_{43}) \boldsymbol{\mathcal{X}} Q_{11} + CON_{\eta_b}, \tag{4.28}$$

where

$$W_{11} = \begin{bmatrix} 1 & 0 & 0 & 0 & 0 \\ 0 & 0 & 0 & 0 & 0 \\ 0 & 0 & 0 & 0 & 0 \\ 0 & 0 & 0 & 0 & 0 \end{bmatrix}, I_{11} = \begin{bmatrix} 1 & 0 & 0 & 0 \\ 0 & 0 & 0 & 0 \\ 0 & 0 & 0 & 0 \\ 0 & 0 & 0 & 0 \end{bmatrix},$$

$$I_{22} = \begin{bmatrix} 0 & 0 & 0 & 0 \\ 0 & 1 & 0 & 0 \\ 0 & 0 & 0 & 0 \\ 0 & 0 & 0 & 0 \end{bmatrix}, CON_{\eta_b} = \begin{bmatrix} 0 & 0 & 0 & 0 \\ 0 & 0 & 0 & 0 \\ 0 & 0 & 0 & 0 \\ \eta_b & 0 & 0 & 0 \end{bmatrix},$$

and $\gamma_1 = 1 - \eta_{pgu}$, $\gamma_2 = h_{411}/\eta_h$, and $\gamma_3 = h_{412}/\eta_h$. Then (4.24) and (4.26) construct the non-linear inequality constraints of the optimization problem.

### 4.3.5    Optimization Algorithm

In the optimization problem, the hourly objective function is a linear one, however, one non-linear equality and fifteen non-linear inequality constrains are involved. This type of problem can be solved by using a variety of methods, for example, penalty and barrier function methods, gradient projection methods, and SQP methods. Among these methods, SQP algorithms have proved highly effective for solving general constrained problems with smooth objective and constraint functions [17]. In this chapter, we adopt the line search method to enable two SQP algorithms to converge with arbitrary initial points; moreover, the Hessians of the Lagrangian are approximated using the Broyden–Fletcher–Goldfarb–Shanno (BFGS) formula. Even though the line search method allows arbitrary initial points, to accelerate the convergence, we manually designate some feasible initial points calculated from FEL, FTL, and other data combinations. By doing so, time for convergence is significantly reduced and the local optimal solution can be avoided. The comparison of different algorithms is beyond the scope of this chapter.

To solve the formulated optimization problem, the optimization toolbox in MAT-LAB is utilized. The optimization function chosen here is *fmincon.m*, whose algorithm is set to be "sqp" (SQP). The computer used to solve this problem is configured with a 2.4 GHz Intel Core 2 Duo processor and 4 GB 1067 MHz memory. In the simulation, with appropriate initial points selected, in order to solve the hourly optimization problem to obtain the optimal power flow in the next hour, the time elapsed is only 0.034 s. This time is much shorter than 1 h, that is, the optimal power flow and operation strategy for the next hour can be readily obtained at the very beginning of this hour. Thus, we can tell that this method is quite efficient. In the off-line optimization for the optimal PGU capacity, the time consumed was 6231.12 s. Even though this time is much longer, it will not affect the whole efficiency of the system, for the off-line optimization will only be processed once.

To obtain the optimal PGU capacity, another step is to minimize the $EC_{annual}$ function. Since the optimizer $\mathcal{X}$ is optimized hourly, under a certain PGU capacity, the system is annually optimized. Thus, we only need to sum a whole year $EC_{hour}$ value to obtain the hourly based optimal $EC_{annual}$ and then search for the optimal PGU capacity $F_{pgumopt}$, which minimizes the $EC_{annual}$ function.

## 4.4    Case Study

### 4.4.1    Hypothetical Building Configuration

The hypothetical building configuration can be referred to in Section 3.4.1.

## 4.4.2  Simulation Parameters

Given the proposed power flow and operation strategy, and other facilities' capacities, the task is to find an optimal PGU capacity. Table 4.1 shows related coefficients in the CCHP system. The fuel chosen here is the natural gas which has been widely used in North America. The weights of the EC are chosen to satisfy different requirements, which implies that we can increase $\omega_3$ to reduce the GHG emissions or increase $\omega_1$ to reduce PEC. In this case study, we set the weights as stated in Table 4.1 to first save the primary energy and then secondly reduce the HTC. The GHG emissions is placed last.

The initial facilities' capacities for optimization, except the PGU capacity, which is the one to be optimized, are chosen according to the simulation data. The initial boilers in the SP system and CCHP system are chosen to be 200 kW and 120 kW,

**Table 4.1**  System coefficients

| Symbol | Variable | Value |
|--------|----------|-------|
| $\eta_{pgu}$ | Efficiency of PGU of CCHP system | 0.3 |
| $\eta_e^{SP}$ | Generation efficiency SP system | 0.35 |
| $\eta_h$ | Efficiency of heating unit | 0.8 |
| $\eta_b$ | Efficiency of boiler | 0.8 |
| $\eta_{hrs}$ | Efficiency of heat recovery system | 0.8 |
| $COP_{ac}$ | Coefficient of performance of absorption chiller | 0.7 |
| $COP_{ec}$ | Coefficient of performance of electric chiller | 3 |
| $\eta_{grid}$ | Transmission efficiency of local grid | 0.92 |
| $\mu_e$ | $CO_2$ emission conversion factor of electricity (g/kWh) | 968 |
| $\mu_f$ | $CO_2$ emission conversion factor of natural gas (g/kWh) | 220 |
| $C_e$ | Electricity rates ($/kWh) | 0.0987 |
| $C_f$ | Natural gas rates ($/kWh) | 0.0577 |
| $C_{ca}$ | Carbon tax rates ($/g) | 0.00003 |
| $C_{pgu}$ | Unit price of PGU ($/kWh) | 1046 |
| $C_b$ | Unit price of boiler ($/kWh) | 46 |
| $C_h$ | Unit price of heating unit ($/kWh) | 30 |
| $C_{ac}$ | Unit price of absorption chiller ($/kWh) | 185 |
| $C_{ec}$ | Unit price of electric chiller ($/kWh) | 149 |
| $L$ | Facilities' lives (year) | 10 |
| $\omega_1$ | Coefficient of PES | 0.5 |
| $\omega_2$ | Coefficient of HTCS | 0.4 |
| $\omega_3$ | Coefficient of CDER | 0.1 |

respectively; the initial heating unit is chosen to be 120 kW; the initial absorption and electric chillers are both chosen to be 120 kW.

### 4.4.3   Test Results

The comparison of FEL, FTL, and the proposed optimal operation strategy is shown in Figures 4.1, 4.2, and 4.3, which are in summer, winter, and spring, respectively.

It can be noticed that, from Figures 4.1, 4.2 and 4.3, the $EC_{hour}$ value of the proposed optimal power flow and operation strategy is no more than those of the FEL and FTL strategies. In summer and winter, especially in the early morning and late night, compared with the FEL and FTL strategies, the proposed optimal power flow or strategy can perform much better. However, in the middle of a day, the proposed optimal strategy can perform slightly better than those two. In spring, the similar case as in autumn, the performance of the proposed optimal strategy is slightly better than the FTL strategy, which is much better than the FEL strategy.

Different from the proposed power flow and operation strategy, both FEL and FTL strategies will inherently waste a certain amount of energy. In FEL, the CCHP system satisfies the electric demands first. If the thermal energy provided by the PGU is not enough, then additional fuel should be purchased to feed the auxiliary boiler to compensate for the thermal gap; if the thermal energy provided exceeds the thermal demands, this part of the thermal energy will be wasted. The same situation exists in the FTL. However, we can tell from the above analyses that the proposed optimal solution can manage the energy input and power flow reasonably to eliminate the waste of energy. As presented in (4.20), the equality constrains the output of the CCHP system to exactly match the demand of the building users. Hence, compared

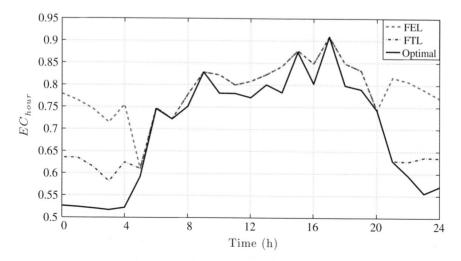

**Figure 4.1**   Comparison of three strategies in a summer day

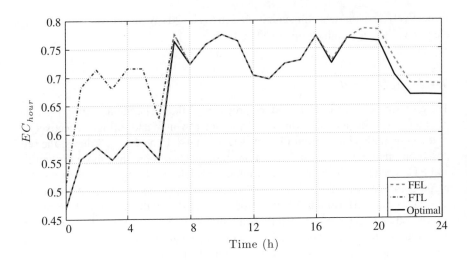

**Figure 4.2**   Comparison of three strategies in a winter day

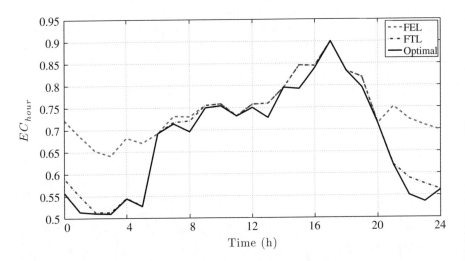

**Figure 4.3**   Comparison of three strategies in a spring day

with the FEL and FTL, with no energy wasted, the proposed optimal strategy can perform better. On the other hand, through optimizing the power flow inside the CCHP
system, the electric cooling to cool load ratio can be adjusted hourly to minimize
the objective function. The above two advantages make the proposed power flow and
operation strategy outperform FEL and FTL, and be the optimal one.

The searching result for the optimal PGU capacity is shown in Figure 4.4. It can
be readily obtained from Figure 4.4 that the optimal PGU capacity is 30 kW. One

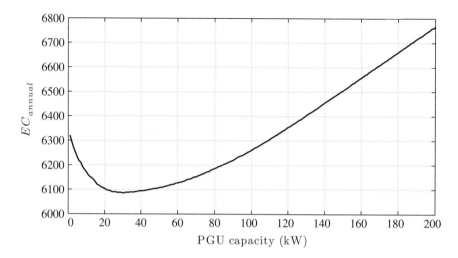

**Figure 4.4**    $EC_{annual}$ of PGU capacity from 0 to 200 kW

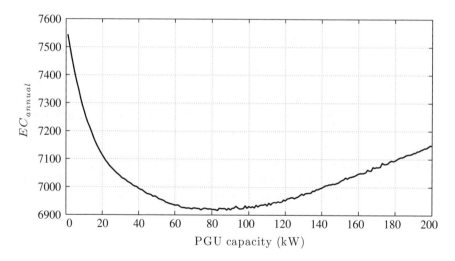

**Figure 4.5**    $EC_{annual}$ of PGU capacity from 0 to 200 kW

of the reasons for the $EC_{annual}$ growing rapidly after 30 kW is the price of the PGU facility. Since the weight for the HTC, say $\omega_2$, is set to be 0.4, the price of the PGU is a much stronger factor in the final objective function. If we decrease the weight for the HTCS to 0.2 and increase the weight for CDER to 0.3, the optimal PGU capacity can be larger, which is shown in Figure 4.5. In a nutshell, the designed optimal operation strategy is dependent on different requirements. The more attention paid to the economical aspect than to the environmental one, the smaller the PGU capacity

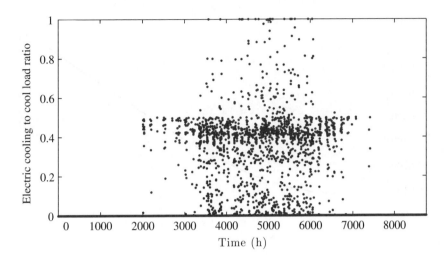

**Figure 4.6**    Variation of the electric cooling to cool load ratio

will be. The trade-off between the cost and the environment should be taken into consideration when designing the operation strategy.

As mentioned above, in order to minimize the objective function in every hour, the electric cooling to cool load ratio must vary according to different energy demand. The variation of this ratio has also been fully considered in the proposed optimal power flow and operation strategy. Figure 4.6 shows the optimal electric cooling to cool load ratio in every hour during 1 year. We can see from this figure that, during 1 year, most of the ratio values vary between 0 and 1 to keep an optimal system performance; some values stay at 0 to make full use of the absorption chiller except for the winter cases; only a few values lie at 1 to take advantage of the high COP of the electric chiller. We can also draw the conclusion that, despite the high COP, the high rate of electricity is still a restriction for the electric chiller.

## 4.5   Summary

A comprehensive yet intuitive approach of matrix modeling for the CCHP system with hybrid chillers has been proposed in this chapter. Efficiency matrices of system components, the conversion matrix of the whole system and dispatch factors are incorporated into this model. The formulation of the CCHP system optimization consists of two parts: (1) an EC function is proposed; and (2) the supply–demand balance is formulated as a non-linear equality constraint, and the output upper and lower bound of the system components are formulated as the inequality constraints. The SQP algorithm has been employed to optimize the EC function. The electric cooling to cool load ratio varies between 0 and 1 to minimize the objective function in every hour. It is worth mentioning that the PEC, HTC, ATC, and the GHG emissions are all considered in the objective function. An illustrative case study is conducted to show the effectiveness and the economic efficiency of the proposed optimal power flow and operation strategy. Compared with FEL and FTL, the proposed operation strategy performs much better in the 1-year simulation test.

## Appendix 4.A    Non-convex Optimization Algorithm

In this appendix, we consider a CCHP system which is allowed to sell the electricity back to the grid. The optimal power flow is designed as a non-convex programming problem. By using the exact penalty formulation, the non-convex programming problem can be solved iteratively in a convex way.

### 4.A.1    EC Function Construction

In this sold-back-enabled CCHP system, PES, HTCS, and CDER are still adopted as the evaluation criteria. However, the definitions of those criteria change accordingly due to the sold-back feature.

#### 4.A.1.1    PES

Taking the electricity sold-back into consideration, PES is defined as

$$PES \triangleq 1 - \frac{F^{CCHP} - E_s}{F^{SP}}, \tag{4.A.1}$$

where $F^{CCHP}$ is the primary energy consumed by the CCHP system and $E_s$ is the sold-back electricity.

#### 4.A.1.2    HTCS

The value of HTC with electricity sold-back considered can be calculated as

$$HTC \triangleq E_{grid}C_e + E_{grid}\mu_e C_{ca} + F_m C_f + F_m \mu_f C_{ca} + \frac{\sum_{k=1}^{l} N_k C_k}{8760 \cdot L}$$
$$- E_s C_s - E_s \mu_e C_{ca}, \tag{4.A.2}$$

where $N_k$ and $C_k$ are the installed capacity of equipments and the initial capital cost of each piece of equipment, respectively; $l$ is the number of pieces of equipment being used and $L$ is the life of a piece of equipment; $C_e$, $C_f$, and $C_s$ are the unit purchasing prices of electricity and the fuel, and unit electricity sold-back price, respectively; $C_{ca}$ is the carbon tax rate, and $\mu_e$ and $\mu_f$ are the carbon conversion factor of the electricity and fuel, respectively. Then the HTCS can be defined as

$$HTCS \triangleq 1 - \frac{HTC^{CCHP}}{HTC^{SP}}. \tag{4.A.3}$$

#### 4.A.1.3 CDER

Considering the electricity sold-back, the total CDE of an energy system can be calculated as

$$CDE = E_{grid}\mu_e + F_m\mu_f - E_s\mu_e, \tag{4.A.4}$$

where $\mu_e$ and $\mu_f$ are the carbon conversion factor of electricity and the fuel, respectively. Then the CDER can be defined as

$$CDER \triangleq 1 - \frac{CDE^{CCHP}}{CDE^{SP}}. \tag{4.A.5}$$

#### 4.A.1.4 EC Function

To measure the performance of the CCHP system, we need to consider all of these three criteria. Thus, we adopt the weighted summation of *PES*, *HTCS*, and *CDER* in the hourly EC function which is defined as

$$EC_{hour} \triangleq 1 - (\omega_1 PES + \omega_2 HTCS + \omega_3 CDER), \tag{4.A.6}$$

where $\omega_1$, $\omega_2$, and $\omega_3$ are the weights of *PES*, *HTCS*, and *CDER*, respectively. The boundary condition is $0 \leq \omega_1, \omega_2, \omega_3 \leq 1$ and $\omega_1 + \omega_2 + \omega_3 = 1$.

### 4.A.2 Optimization Problem Formulation

#### 4.A.2.1 Objective Function

Having the optimizer $\mathcal{X}$ defined as in (4.17) and considering the electricity sold-back feature, the $EC_{hour}$ function can be rewritten as a non-convex function of the optimizer $\mathcal{X}$ as

$$EC_{hour}(\mathcal{X}) = \omega_1\frac{P\mathcal{X}}{F^{SP}} + \omega_2\frac{C\mathcal{X} + \mathcal{L}}{HTC^{SP}} + \omega_3\frac{D\mathcal{X}}{CDE^{SP}} + \omega_1\frac{P_s e_2^{\mathsf{T}}(\mathcal{V}_o - \mathcal{H}P\mathcal{X})}{F^{SP}}$$
$$+ \omega_2\frac{C_s e_2^{\mathsf{T}}(\mathcal{V}_o - \mathcal{H}P\mathcal{X})}{HTC^{SP}} + \omega_3\frac{D_s e_2^{\mathsf{T}}(\mathcal{V}_o - \mathcal{H}P\mathcal{X})}{CDE^{SP}} \tag{4.A.7}$$

where
$$P = \begin{bmatrix} 0 & 0 & 0 & 1 & 1/(\eta_e^{SP}\eta_{grid}) \end{bmatrix},$$
$$P_s = 1/(\eta_e^{SP}\eta_{grid}),$$
$$C = \begin{bmatrix} 0 & 0 & 0 & C_f + \mu_f C_c & C_e + \mu_e C_c \end{bmatrix},$$
$$C_s = C_s + \mu_e C_c,$$
$$D = \begin{bmatrix} 0 & 0 & 0 & \mu_f & \mu_e \end{bmatrix},$$
$$D_s = \mu_e,$$
$$\mathcal{L} = \frac{\sum_{k=1}^{l} N_k C_k}{8760 \cdot L},$$

and $\mathbf{e}_2 = [0\ 1\ 0\ 0]^T$ is the standard basis vector. In this book, $\mathbf{e}_i$ denotes the standard basis vector with the $i$th element being 1.

Then the hourly optimization problem becomes

$$\underbrace{\min_{\mathfrak{x}}}\ \{EC_{hour}(\mathfrak{X})\}, \tag{4.A.8}$$

and is subject to non-linear equality and inequality constraints, leading to a non-convex optimization problem. This objective function is a comprehensive one, which integrates three aspects, that is, PES, HTCS, and CDER, into it. Three weights, $\omega_1$, $\omega_2$, and $\omega_3$, are selected according to different requirements. For example, if more emphasis needs to be put on the CDER, then $\omega_3$ should be raised and $\omega_1$ and $\omega_2$ should be accordingly decreased.

### 4.A.2.2   Non-convex and Non-linear Equality Constraints

The non-convex and non-linear equality constraint in (4.20) can be rewritten as

$$\mathbf{e}_3^T(\mathcal{V}_o - \mathcal{H}\mathcal{V}_i) = 0, \tag{4.A.9}$$

$$\mathbf{e}_4^T(\mathcal{V}_o - \mathcal{H}\mathcal{V}_i) = 0. \tag{4.A.10}$$

Similarly, there is no need to constrain the supply and demand relation for each component, for this type of constraint is already included in the conversion matrix $\mathcal{H}$.

Having the system input $\mathcal{V}_i$ written as in (4.22), (4.A.9) and (4.A.10) can be represented as

$$a_3(\mathfrak{X}) = \mathbf{e}_3^T\mathcal{V}_o - \mathbf{e}_3^T(\mathcal{H}_1\mathfrak{X}\mathfrak{X}^T U_{11} P\mathfrak{X} + \mathcal{H}_2\mathfrak{X}Q_{11}P\mathfrak{X} + \mathcal{H}_3\mathfrak{X}Q_{12}P\mathfrak{X} + \mathcal{H}_4P\mathfrak{X})$$
$$= 0, \tag{4.A.11a}$$

$$a_4(\mathfrak{X}) = \mathbf{e}_4^T\mathcal{V}_o - \mathbf{e}_4^T(\mathcal{H}_1\mathfrak{X}\mathfrak{X}^T U_{11} P\mathfrak{X} + \mathcal{H}_2\mathfrak{X}Q_{11}P\mathfrak{X} + \mathcal{H}_3\mathfrak{X}Q_{12}P\mathfrak{X} + \mathcal{H}_4P\mathfrak{X})$$
$$= 0. \tag{4.A.11b}$$

### 4.A.2.3   Linear Inequality Constraints

To make all constraints in the standard form, inequalities in (4.24) are rewritten as

$$c_\alpha(\mathfrak{X}) = -\alpha \le 0, \tag{4.A.12a}$$

$$c_{\alpha-}(\mathfrak{X}) = \alpha - 1 \le 0, \tag{4.A.12b}$$

$$c_\beta(\mathfrak{X}) = -\beta \le 0. \tag{4.A.12c}$$

#### 4.A.2.4   Non-convex and Non-linear Inequality Constraints

The output upper and lower bounds constraints in (4.27) can be written in the standard form as

$$\overline{c}^{\ell}(\mathcal{X}) = \mathcal{H}^{\ell}\Gamma_{\ell}P\mathcal{X} - \overline{\mathcal{V}}^{\ell} \leq 0, \tag{4.A.13a}$$

$$\underline{c}^{\ell}(\mathcal{X}) = \underline{\mathcal{V}}^{\ell} - \mathcal{H}^{\ell}\Gamma_{\ell}P\mathcal{X} \leq 0. \tag{4.A.13b}$$

In addition to (4.A.12) and (4.A.13), the electricity sold-back feature also introduces an inequality constraint, which can be represented as

$$c_e(\mathcal{X}) = e_2^{\mathsf{T}}(\mathcal{V}_o - \mathcal{H}P\mathcal{X}) \leq 0. \tag{4.A.14}$$

Then (4.A.12), (4.A.13), and (4.A.14) construct the non-convex inequality constraints of the optimization problem in (4.A.8).

### 4.A.3   Optimization Algorithm

Since the objective function, equality constraints, and most of the inequality constraints are non-convex, this optimization problem is a non-convex programming problem. One issue involved in the non-convex programming problem is that the global solution may not be found. However, in the engineering problem, quickly finding out a good local optimal solution is way more important than finding the global solution. In the following, we would like to use some techniques to find a good enough local solution that is close to the global solution.

#### 4.A.3.1   An Exact Penalty Formulation

There exist many approaches to handle the non-convex programming problem [17]. Among those, one popular approach is the exact penalty formulation. The exact penalty formulation is a modified version of the sequential convex programming (SCP), which is used to handle the non-convex programming problem. Suppose we have the non-convex programming problem represented as

$$\min f(\mathcal{X})$$

$$\text{s.t. } a_i(\mathcal{X}) = 0, \text{ for } i = 1, 2, \dots, p, \tag{4.A.15}$$

$$c_j(\mathcal{X}) \leq 0 \text{ , for } j = 1, 2, \dots, q,$$

where $f(\mathcal{X})$ and $c_j(\mathcal{X})$ may not be convex and $a_i(\mathcal{X})$ may not be affine. Different from SCP, the exact penalty formulation does not suffer the infeasibility issue. Formulated by the exact penalty technique, we have (4.A.15) modified as

$$\min F_{\tau}(\mathcal{X}) = f(\mathcal{X}) + \tau \left( \sum_{i=1}^{p} |a_i(\mathcal{X})| + \sum_{j=1}^{q} c_j(\mathcal{X})_+ \right), \tag{4.A.16}$$

where $c_j(\mathbf{X})_+ = \max\{c_j(\mathbf{X}), 0\}$ and $\tau > 0$ is sufficiently large. Then solving (4.A.16) becomes solving a sequential convex problem

$$\min \ \tilde{F}_\tau(\mathbf{X}) = \tilde{f}(\mathbf{X}) + \tau \left( \sum_{i=1}^{p} |\tilde{a}_i(\mathbf{X})| + \sum_{j=1}^{q} \tilde{c}_j(\mathbf{X})_+ \right),$$

$$\text{s.t. } \mathbf{X} \in \mathcal{R}^k,$$

(4.A.17)

where $\tilde{f}(\mathbf{X})$, $\tilde{a}_i(\mathbf{X})$, and $\tilde{c}_j(\mathbf{X})$ are the convex approximations of $f(\mathbf{X})$, $a_i(\mathbf{X})$, and $c_j(\mathbf{X})$, respectively. $\mathcal{R}^k$ is the trust region for the $k$th iteration, where the approximations are carried out. To summarize, for the $k$th iteration, solving (4.A.15) can be divided into the following steps:

**Step 1:** Convexify $f(\mathbf{X})$ and $c_j(\mathbf{X})$ around an initial point $\mathbf{X}_{k0}$ to obtain $\tilde{f}(\mathbf{X})$ and $\tilde{c}_j(\mathbf{X})$, respectively.
**Step 2:** Linearize $a_i(\mathbf{X})$ around the initial point $\mathbf{X}_{k0}$ to obtain $\tilde{a}_i(\mathbf{X})$.
**Step 3:** Solve the convex programming problem in (4.A.17) in the trust region $\mathcal{R}^k$ to obtain the optimal solution $\mathbf{X}_k^*$ for the $k$th iteration. If $|f(\mathbf{X}_k^*) - f(\mathbf{X}_{k0})| \leq \epsilon$ or $k = K$, output $\mathbf{X}_k^*$, otherwise, go back to Step 1 and use $\mathbf{X}_k^*$ as the initial point for the next iteration.

**Remark 4.A.1:** In SCP, if the trust region is not appropriately selected, the optimization problem may be infeasible. However, if we choose the exact penalty formulation, the optimization problem is always feasible. Note that, if both the SCP and exact penalty formulated problem are feasible, the convergence rate of the former is faster than the latter; and the equality constraints of the latter are not always satisfied but slightly away from 0. In practice, for many engineering problems, infeasibility is an issue way more important than a non-perfect equality constraint.

### 4.A.3.1.1   *Convexification of the Objective Function*
In the following, we use $f(\mathbf{X})$ to denote the the original non-convex objective function (4.A.7). Since in (4.A.7), only the $\mathbf{e}_2^\mathsf{T} \mathcal{H} P \mathbf{X}$ is non-convex, we use Taylor expansion to convexify $f(\mathbf{X})$. Let

$$\hat{f}_i(\mathbf{X}) = \mathbf{e}_i^\mathsf{T} \mathcal{H} P \mathbf{X}$$
$$= \mathbf{e}_i^\mathsf{T} \mathcal{H}_1 \mathbf{X} \mathbf{X}^\mathsf{T} U_{11} P \mathbf{X} + \mathbf{e}_i^\mathsf{T} \mathcal{H}_2 \mathbf{X} Q_{11} P \mathbf{X} + \mathbf{e}_i^\mathsf{T} \mathcal{H}_3 \mathbf{X} Q_{12} P \mathbf{X} + \mathbf{e}_i^\mathsf{T} \mathcal{H}_4 P \mathbf{X},$$

(4.A.18)

denote the $i$th element of $\hat{f}(\mathbf{X}) = \mathcal{H} P \mathbf{X}$.
Consequently we can have the gradient calculated as

$$\nabla \hat{f}_i(\mathbf{X}) = \mathbf{e}_i^\mathsf{T} \mathcal{H}_1 \mathbf{X} (U_{11} P \mathbf{X} + P^\mathsf{T} U_{11}^\mathsf{T} \mathbf{X}) + \mathbf{X}^\mathsf{T} U_{11} P \mathbf{X} \mathcal{H}_1^\mathsf{T} \mathbf{e}_i$$
$$+ \mathbf{e}_i^\mathsf{T} \mathcal{H}_2 \mathbf{X} P^\mathsf{T} Q_{11}^\mathsf{T} + Q_{11} P \mathbf{X} \mathcal{H}_2^\mathsf{T} \mathbf{e}_i$$
$$+ \mathbf{e}_i^\mathsf{T} \mathcal{H}_3 \mathbf{X} P^\mathsf{T} Q_{12}^\mathsf{T} + Q_{12} P \mathbf{X} \mathcal{H}_3^\mathsf{T} \mathbf{e}_i$$
$$+ \mathcal{H}_4^\mathsf{T} \mathbf{e}_i.$$

(4.A.19)

Thus, in a sufficiently small region, that is, $\| \boldsymbol{\mathcal{X}} - \boldsymbol{\mathcal{X}}_{k0} \|_2 \leq \epsilon, \hat{f}_i(\boldsymbol{\mathcal{X}})$ can be represented as

$$\hat{f}_i(\boldsymbol{\mathcal{X}}) = \hat{f}_i(\boldsymbol{\mathcal{X}}_0) + \nabla^\mathsf{T} \hat{f}_i(\boldsymbol{\mathcal{X}}_0)(\boldsymbol{\mathcal{X}} - \boldsymbol{\mathcal{X}}_{k0}) + o. \qquad (4.A.20)$$

As a result, we can write the objective function around $\boldsymbol{\mathcal{X}}_{k0}$ as

$$
\begin{aligned}
\tilde{f}(\boldsymbol{\mathcal{X}}) = {}& \omega_1 \frac{P\boldsymbol{\mathcal{X}}}{F^{SP}} + \omega_2 \frac{C\boldsymbol{\mathcal{X}} + \mathcal{L}}{HTC^{SP}} + \omega_3 \frac{D\boldsymbol{\mathcal{X}}}{CDE^{SP}} \\
& - \left( \frac{\omega_1 P_s}{F^{SP}} + \frac{\omega_2 C_s}{HTC^{SP}} + \frac{\omega_3 D_s}{CDE^{SP}} \right) \nabla^\mathsf{T} \hat{f}_2(\boldsymbol{\mathcal{X}}_{k0})(\boldsymbol{\mathcal{X}} - \boldsymbol{\mathcal{X}}_{k0}) \qquad (4.A.21) \\
& + \left( \frac{\omega_1 P_s}{F^{SP}} + \frac{\omega_2 C_s}{HTC^{SP}} + \frac{\omega_3 D_s}{CDE^{SP}} \right) (\mathbf{e}_2^\mathsf{T} \mathcal{V}_o - \hat{f}_2(\boldsymbol{\mathcal{X}}_{k0})),
\end{aligned}
$$

which is a convex function.

#### 4.A.3.1.2   Convexification of the Non-convex Equality Constraints

To convexify the non-convex equality constraints, we shall still use the same technique as for dealing with the objective function, that is, Taylor expansion. By following similar lines as in Section 4.A.3.1.1, (4.23) becomes

$$
\begin{aligned}
\tilde{a}_3(\boldsymbol{\mathcal{X}}) &= - \nabla^\mathsf{T} \hat{f}_3(\boldsymbol{\mathcal{X}}_0)(\boldsymbol{\mathcal{X}} - \boldsymbol{\mathcal{X}}_0) + \mathbf{e}_3^\mathsf{T} \mathcal{V}_o - \hat{f}_3(\boldsymbol{\mathcal{X}}_0) = 0, \\
\tilde{a}_4(\boldsymbol{\mathcal{X}}) &= - \nabla^\mathsf{T} \hat{f}_4(\boldsymbol{\mathcal{X}}_0)(\boldsymbol{\mathcal{X}} - \boldsymbol{\mathcal{X}}_0) + \mathbf{e}_4^\mathsf{T} \mathcal{V}_o - \hat{f}_4(\boldsymbol{\mathcal{X}}_0) = 0,
\end{aligned}
\qquad (4.A.22)
$$

which are both affine.

### 4.A.3.2   Convexification of the Non-convex Inequality Constraints

Constraints (4.A.12) are all linear, they will not trouble us when solving the optimization problem. For constraints (4.A.13), since we have the PGU, boiler, heat recovery system, electric chiller, absorption chiller, and heating unit, the total number of non-convex inequality constraints is twelve. Componentwisely, each non-convex inequality contains four sub-constraints. Hence, in total, we have forty eight non-convex constraints.

**Remark 4.A.2:** In order to improve the algorithm efficiency, some redundant constraints can be removed. For example, in the upper bound constraint of the PGU, only electricity and heating generation need to be bounded, as a result, the constraints corresponding to fuel and cooling generation can be removed. Consequently, the original forty eight non-convex inequality constraints are reduced to fourteen non-convex inequality constraints. ∎

In the following, we shall take a general unit upper bound as an example to illustrate the convexification method of the non-convex inequality constraints. From (4.A.13), componentwisely, we have the above mentioned inequality represented as

$$\mathbf{e}_j^\mathsf{T} (\mathcal{H}^\ell \Gamma_\ell P \boldsymbol{\mathcal{X}} - \overline{\mathcal{V}}^\ell) \leq 0. \qquad (4.A.23)$$

We shall use the data interpolation by convex functions technique to convexify (4.A.23). A useful Lemma will be introduced.

**Lemma 4.A.1:**  There exists a convex function $f(x)$ with dom($f$) $= \mathcal{R}^n$ that satisfies $f(x_i) = y_i$ for $i = 1, 2, \ldots, m$, if and only if there exist vectors $\mathbf{g}_1, \mathbf{g}_2, \ldots, \mathbf{g}_m$ such that

$$y_j \geq y_i + \mathbf{g}_i^\mathsf{T}(x_j - x_i), \quad \text{for } i,j = 1, \ldots, m. \tag{4.A.24}$$

Then a convex function $f(x)$ can be constructed as

$$f(x) = \max_{i=1,2,\ldots,m} \{y_i + \mathbf{g}_i^\mathsf{T}(x - x_i)\}. \tag{4.A.25}$$

If the data $\{(x_i, \hat{y}_i), i = 1, 2, \ldots, m\}$ are given, then the problem of fitting a convex function can be formulated as

$$\min \sum_{i=1}^{m} (y_i - \hat{y}_i)^2 \tag{4.A.26}$$

$$\text{s.t. } y_i - y_j - \mathbf{g}_i^\mathsf{T}(x_i - x_j) \leq 0 \quad \text{for } i,j = 1, 2, \ldots, m,$$

where $y_i$ and $\mathbf{g}_i$ are treated as variables.   ∎

We first prove that the optimization problem in **Lemma 4.A.1** can be formulated as a convex QP problem in the form of

$$\min f(x) = \frac{1}{2}\mathbf{x}^\mathsf{T}\mathbf{H}\mathbf{x} + \mathbf{x}^\mathsf{T}\mathbf{p} \tag{4.A.27}$$

$$\text{s.t. } \mathbf{A}\mathbf{x} \leq \mathbf{b}.$$

**Proof.**  We can define variables

$$\mathbf{y} = [y_1 \ y_2 \ \cdots \ y_m \ \mathbf{g}_1^\mathsf{T} \ \mathbf{g}_2^\mathsf{T} \ \cdots \ \mathbf{g}_m^\mathsf{T}]^\mathsf{T}, \tag{4.A.28}$$

which is to be optimized, and

$$\hat{\mathbf{y}} = [\hat{y}_1 \ \hat{y}_2 \ \cdots \ \hat{y}_m]^\mathsf{T},$$
$$\mathbf{x} = [\mathbf{x}_1^\mathsf{T} \ \mathbf{x}_2^\mathsf{T} \ \cdots \ \mathbf{x}_m^\mathsf{T}]^\mathsf{T}, \tag{4.A.29}$$

which are constants. For the convenience of the following derivation, we define

$$\mathbf{Y}_1 = [\mathbf{I}_m \ 0 \ \cdots \ 0],$$

$$\mathbf{Y}_{i+1} = [0 \ \cdots \ 0 \ \underbrace{\mathbf{I}_n}_{i+1th \ block} \ 0 \ \cdots \ 0],$$

$$\mathbf{D}_{ijy} = [0 \ \cdots \ 0 \ \underbrace{1}_{ith \ element} \ 0 \ \cdots \ 0 \ \underbrace{-1}_{jth \ element} \ 0 \ \cdots \ 0], \tag{4.A.30}$$

$$\mathbf{D}_{ijx} = [0 \ \cdots \ 0 \ \underbrace{\mathbf{I}_n}_{ith \ block} \ 0 \ \cdots \ 0 \ \underbrace{-\mathbf{I}_n}_{jth \ block} \ 0 \ \cdots \ 0].$$

The objective function $f(y)$ can be rewritten as

$$f(\mathbf{y}) = \sum_{i=1}^{m}(y_i^2 - 2\hat{y}_i y_i + \hat{y}_i^2)$$

$$= \sum_{i=1}^{m} y_i^2 - \sum_{i=1}^{m} 2\hat{y}_i y_i + \sum_{i=1}^{m} \hat{y}_i^2 \qquad (4.A.31)$$

$$\Longleftrightarrow \frac{1}{2}\mathbf{y}^\mathsf{T}\mathbf{H}\mathbf{y} + \mathbf{y}^\mathsf{T}\mathbf{p},$$

where

$$\mathbf{H} = 2\mathbf{Y}_1^\mathsf{T}\mathbf{Y}_1 = \begin{bmatrix} 2\mathbf{I}_m & \mathbf{0} \\ \mathbf{0} & \mathbf{0} \end{bmatrix},$$

$$\qquad (4.A.32)$$

$$\mathbf{p} = -2\mathbf{Y}_1^\mathsf{T}\hat{\mathbf{y}} = \begin{bmatrix} -2\hat{\mathbf{y}} \\ \mathbf{0} \end{bmatrix}.$$

For any $i$ and $j$, the inequality constraint can be represented as

$$y_i - y_j - (\mathbf{x}_i - \mathbf{x}_j)^\mathsf{T}\mathbf{g}_i$$

$$= \mathbf{D}_{ijy}\mathbf{Y}_1\mathbf{y} - (\mathbf{D}_{ijx}\mathbf{x})^\mathsf{T}\mathbf{Y}_{i+1}\mathbf{y}$$

$$= \mathbf{A}_{ij}\mathbf{y} \qquad (4.A.33)$$

$$\leq \mathbf{0},$$

where

$$\mathbf{A}_{ij} = \mathbf{D}_{ijy}\mathbf{Y}_1 - \mathbf{x}^\mathsf{T}\mathbf{D}_{ijx}^\mathsf{T}\mathbf{Y}_{i+1}. \qquad (4.A.34)$$

Thus, in standard form, the original problem can be written as

$$\min f(\mathbf{y}) = \frac{1}{2}\mathbf{y}^\mathsf{T}\mathbf{H}\mathbf{y} + \mathbf{y}^\mathsf{T}\mathbf{p},$$

$$\text{s.t. } \mathbf{A}\mathbf{y} \leq \mathbf{0}, \qquad (4.A.35)$$

where

$$\mathbf{A} = \begin{bmatrix} \mathbf{A}_{12} \\ \mathbf{A}_{13} \\ \vdots \\ \mathbf{A}_{1m} \\ \mathbf{A}_{21} \\ \mathbf{A}_{23} \\ \vdots \\ \mathbf{A}_{2m} \\ \vdots \\ \mathbf{A}_{m(m-1)} \end{bmatrix}. \qquad (4.A.36)$$

Note that the constraint is redundant when $i = j$. Thus, there is no $A_{ii}$ in the matrix $A$.    ∎

If we construct the non-convex function

$$\overline{V}_j^{\ell}(\mathbf{X}) = \mathbf{e}_j^{\mathsf{T}}(H^{\ell}\Gamma_{\ell}P\mathbf{X} - \overline{V}^{\ell}) \tag{4.A.37}$$

then the algorithm of using a convex function to approximate this non-convex function in the $k$th iteration can be described as:

**Step 1:** Given a feasible point $\mathbf{X}_{k0}$ and generate $m$ points around $\mathbf{X}_{k0}$ satisfying $\|\mathbf{X}_{ki} - \mathbf{X}_{k0}\| \leq \sigma$ for $i = 1, \dots, m$.
**Step 2:** Compute $\hat{y}_i = \overline{V}_j^{\ell}(\mathbf{X}_i)$ for $i = 0, \dots, m$.
**Step 3:** Solve (4.A.27) to obtain $y_i$ and $\mathbf{g}_i^{\mathsf{T}}$ for $i = 0, \dots, m$.
**Step 4:** Use

$$\widetilde{\overline{c}}_j^{\ell}(\mathbf{X})$$

$$= \max_{i=1,2,\dots,m} \{y_i + \mathbf{g}_i^{\mathsf{T}}(\mathbf{X} - \mathbf{X}_i)\} \tag{4.A.38}$$

$$\leq 0$$

to approximate $\mathbf{e}_j^{\mathsf{T}}(H^{\ell}\Gamma_{\ell}P\mathbf{X} - \overline{V}^{\ell}) \leq 0$ around $\mathbf{X}_{k0}$.

**Remark 4.A.3:** We can follow a similar procedure to construct $\widetilde{\underline{c}}_j^{\ell}(\mathbf{X}) \leq 0$ to approximate $\mathbf{e}_j^{\mathsf{T}}(\underline{V}^{\ell} - H^{\ell}\Gamma_{\ell}P\mathbf{X}) \leq 0$ around $\mathbf{X}_{k0}$. Both $\widetilde{\overline{c}}_j^{\ell}(\mathbf{X})$ and $\underline{\widetilde{c}}_j^{\ell}(\mathbf{X})$ are convex.    ∎

In addition to the device constraints, the non-convex inequality constraint for including the electricity sold back-feature, that is, (4.A.14), should also be convex-ified. Different from the above mentioned algorithm, we use the Taylor expansion to approximate this constraint, which is similar to the method used to deal with the non-convex equality constraints. By following similar lines to Section 4.A.3.1.2, we can have

$$\tilde{c}_e(\mathbf{X}) = -\nabla^{\mathsf{T}}\hat{f}_2(\mathbf{X}_0)(\mathbf{X} - \mathbf{X}_0) + \mathbf{e}_2^{\mathsf{T}}\mathcal{V}_o - \hat{f}_2(\mathbf{X}_0) \leq 0 \tag{4.A.39}$$

as the convex approximation for (4.A.14).

### 4.A.3.3   Summary

We define a set $\mathcal{U} = \{$PGU, boiler, heat recovery system, electric chiller, absorption chiller, heating unit$\}$ and the domain $\mathcal{R}_{\ell}$ where $j$ should take values from for the device $\ell$. By combining the above, we can have the hourly optimal power flow iteratively solved from

$$\min \tilde{f}(\mathcal{X}) + \tau \left[ \sum_{i=3}^{4} |\tilde{a}_i(\mathcal{X})| + \sum_{\ell \in \mathcal{U}} \sum_{j \in \mathcal{R}_\ell} (\tilde{c}_j^\ell(\mathcal{X})_+ + \underline{\tilde{c}}_j^\ell(\mathcal{X})_+) + \tilde{c}_e(\mathcal{X})_+ \right.$$

$$\left. + c_\alpha(\mathcal{X})_+ + c_{\alpha-}(\mathcal{X})_+ + c_\beta(\mathcal{X})_+ \right]$$

(4.A.40)

$$\text{s.t. } \| \mathbf{e}_i^\mathsf{T}(\mathcal{X} - \mathcal{X}_0) \|_2 \le \sigma_i, \text{ for } i = 1, 2, 3, 4, 5,$$

where $\mathbf{e}_i \in \mathbb{R}^5$.

# References

[1] M. Geidl, G. Koeppel, P. Favre-Perrod, B. Klockl, G. Andersson, and K. Frohlich, "Energy hubs for the future," *IEEE Power & Energy Magzine*, vol. 5, no. 1, pp. 24–30, 2007.

[2] M. Geidl and G. Andersson, "A modeling and optimization approach for multiple energy carrier power flow," in *Proceedings of IEEE Russia Power Tech Conference*, St. Petersburg, Russia, June 27–30, 2005, pp. 1–7.

[3] M. Geidl and G. Andersson, "Optimal power dispatch and conversion in systems with multiple energy carriers," in *Proceedings of Power Systems Computation Conference*, 2005, pp. 1–7.

[4] W. Leontief, *Input–output Economics*. Oxford University Press, 1986.

[5] M. Geidl and G. Andersson, "Optimal power flow of multiple energy carriers," *IEEE Transactions on Power Systems*, vol. 22, no. 1, pp. 145–155, 2007.

[6] G. Chicco and P. Mancarella, "Matrix modelling of small-scale trigeneration systems and application to operational optimization," *Energy*, vol. 34, no. 3, pp. 261–273, 2009.

[7] P. Arcuri, G. Florio, and P. Fragiacomo, "A mixed integer programming model for optimal design of trigeneration in a hospital complex," *Energy*, vol. 32, no. 8, pp. 1430–1447, 2010.

[8] H. Cho, P. J. Mago, R. Luck, and L. M. Chamra, "Evaluation of CCHP systems performance based on operational cost, primary energy consumption, and carbon dioxide emission by utilizing an optimal operation scheme," *Applied Energy*, vol. 86, no. 12, pp. 2540–2549, 2009.

[9] H. Ren, W. Gao, and Y. Ruan, "Optimal sizing for residential CHP system," *Applied Thermal Engineering*, vol. 28, no. 5–6, pp. 514–523, 2008.

[10] A. Rong and R. Lahdelma, "An efficient linear programming model and optimization algorithm for trigeneration," *Applied Energy*, vol. 82, no. 1, pp. 40–63, 2005.

[11] B. Zhang and W. Long, "An optimal sizing method for cogeneration plants," *Energy and Buildings*, vol. 38, no. 3, pp. 189–195, 2006.

[12] D. Ziher and A. Poredos, "Economics of a trigeneration system in a hospital," *Applied Thermal Engineering*, vol. 26, no. 7, pp. 680–687, 2006.

[13] J. Wang, Z. Zhai, Y. Jing, and C. Zhang, "Particle swarm optimization for redundant building cooling heating and power system," *Applied Energy*, vol. 87, no. 12, pp. 3668–3679, 2010.

[14] J. Wang, Y. Jing, and C. Zhang, "Optimization of capacity and operation for CCHP system by genetic algorithm," *Applied Energy*, vol. 87, no. 4, pp. 1325–1335, 2010.

[15] T. Savola, T.-M. Tveit, and C.-J. Fogelholm, "A MINLP model including the pressure levels and multiperiods for CHP process optimisation," *Applied Thermal Engineering*, vol. 27, no. 11–12, pp. 1857–1867, 2007.

[16] T. Savola and C.-J. Fogelholm, "MINLP optimisation model for increased power production in small-scale CHP plants," *Applied Thermal Engineering*, vol. 27, no. 1, pp. 89–99, 2007.

[17] A. Antoniou and W.-S. Lu, *Practical Optimization: Algorithms and Engineering Applications*. Springer, 2007.

# 5

# Short-Term Load Forecasting and Post-Strategy Design for CCHP Systems

## 5.1 Introduction and Related Work

Having a designed operation strategy, another aspect that can restrict the CCHP system performance is the load profile. Almost all of the operation strategies are designed by assuming that accurate loads can be obtained in real time. However, this is not the case in reality. In practical applications, we cannot get full access to the heating, cooling and electrical loads in the following hour. The only information that can be used is the historical loads, current and historical dry-bulb, and dew-point temperature. Thus, the question arises naturally: Can we make use of the limited information to forecast the loads in the following hour? To forecast the load, the first step is to construct a forecasting model. In the literature, there exist many time series forecasting models, for example, AutoRegressive (AR) models, Moving Average (MA) models, AutoRegressive Moving Average (ARMA) models [1, 2, 3], AutoRegressive Integrated Moving Average (ARIMA) models [4], and AutoRegressive Moving Average with eXogenous inputs (ARMAX) models. The comparisons among different time series models can be found in [5, 6, 7]. Load forecasting can also be accomplished by adopting an artificial neural network (ANN) [8, 9] and Kalman filter [7, 10, 11]. Other control approaches related to load forecasting and system identification can be found in [12, 13, 14, 15, 16], to name a few. Here, in this chapter, since the temperature information is available, the ARMAX model will be selected as the forecasting model; and the structure of two-stage least squares

*Combined Cooling, Heating, and Power Systems: Modeling, Optimization, and Operation*, First Edition.
Yang Shi, Mingxi Liu, and Fang Fang.
© 2017 John Wiley & Sons Ltd. Published 2017 by John Wiley & Sons Ltd.

(TSLS) will be chosen to identify the model parameters. In order to improve the forecasting accuracy, instead of using the dry-bulb temperature only, the dew-point temperature will be considered as an instrument variable (IV). By doing so, the correlation between the dry-bulb temperature and the error term can be eliminated, and more weather information, such as the humidity, can be included inherently in the model. Thus, the estimated forecasting model can be more accurate than the one without using IV. The first stage of the parameter identification can be readily accomplished by ordinary least squares (OLS); since the second stage will involve a large amount of data, the two-stage recursive least squares (TSRLS) [17, 18], which has been proven to have a faster convergence rate and more accurate estimation, is adopted to reduce the space, computational, and time complexity. Combining OLS and TSRLS in one TSLS procedure is one of the objectives and contributions of this chapter.

Having the forecasted loads, we can directly use the optimal operation strategy proposed in Chapter 4 to determine the energy input and power flow inside the system. However, all of the energy input and power flow determined at this stage are only for the forecasted loads. There still exist discrepancies between the forecasted value and the actual one. The second objective of this chapter is to design a post-strategy after the system fulfilling the forecasted loads in order to compensate for the inaccuracy of forecasting. The designed post-strategy should be able to appropriately coordinate the power flow to use the surplus of one type of energy to compensate for the shortage of other types of energy.

The contributions of this chapter are two-fold:

- We first propose an enhanced TSLS algorithm, which combines OLS and TSRLS, to identify the load forecasting model. The identification procedure is faster than the conventional TSLS algorithm in terms of the time complexity. The identified model can generate more accurate load forecasting results, compared with benchmark models using the conventional TSLS algorithm and ARMAX model without IV. Note that the accuracy in this chapter represents the accuracy of the forecasted load instead of the estimated model parameters.
- We then propose a new post-strategy to compensate for the discrepancies between the actual loads and forecasted loads. Compared with other existing post-strategies, the proposed one can effectively lower the energy consumption, annual cost, and GHG emissions.

This chapter is organized in the following way. In Section 5.2, the forecasting model and the parameter identification algorithm, including IV, TSRLS, and OLS-TSRLS, will be discussed. Load forecasting is briefly introduced at the end of Section 5.2. Section 5.3 investigates two steps of the operation strategy design, including the optimal operation strategy for forecasted loads and the post-strategy. A case study is conducted in Section 5.4 to verify the feasibility and effectiveness of the proposed forecasting method and post-strategy. Section 5.5 concludes this chapter. In the appendix, derivations of the closed-form solutions of quadratic ARMA and ARMAX model identifications are provided.

## 5.2　Estimation Model and Load Forecasting

In this section, an ARMAX model for forecasting the cooling, heating and electrical loads will be proposed first. Algorithms for identifying the model parameters will then be introduced. Since the model for heating $y_h(t)$, cooling $y_c(t)$ and electrical $y_e(t)$ loads are the same, yet with different parameters, only the heating load is taken for example to illustrate the identification method. The model for forecasting the heating load $y_h(t)$ can be expressed as

$$\mathfrak{A}(z)y_h(t) = \alpha_0 + \mathfrak{B}(z)T_{dry}(t) + \mathfrak{C}(z)\epsilon(t), \tag{5.1}$$

where

$$\mathfrak{A}(z) = 1 + \alpha_1 z^{-1} + \alpha_2 z^{-2} + \cdots + \alpha_{n_\alpha} z^{-n_\alpha},$$

$$\mathfrak{B}(z) = \beta_0 + \beta_1 z^{-1} + \beta_2 z^{-2} + \cdots + \beta_{n_\beta} z^{-n_\beta}, \tag{5.2}$$

$$\mathfrak{C}(z) = 1 + \pi_1 z^{-1} + \pi_2 z^{-2} + \cdots + \pi_{n_\pi} z^{-n_\pi},$$

and $\alpha_0$ is a constant term. In (5.1) and (5.2), $T_{dry}(t)$ and $\epsilon(t)$ represent dry-bulb temperature and error term at time $t$, respectively. $y_h(t)$ represents the heating load during the time interval $[t, t+1]$. $z^{-*}$ denotes $*$ time lags from the time instant $t$. In (5.2), $\alpha_i$ ($i = 1, 2, \ldots, n_\alpha$), $\beta_j$ ($j = 0, 1, 2, \ldots, n_\beta$), and $\pi_k$ ($k = 1, 2, \ldots, n_\pi$) represent the model parameters. The objective is to forecast $y_h(t)$ by using the historical heating loads, current/historical dry-bulb temperature, and the error term.

The parameter identification algorithm is selected to be TSLS. In the first stage, we choose an IV and then use OLS to obtain the IV estimation of the explanatory variable $T_{dry}(t)$. In the second stage, TSRLS is adopted to identify the parameters in (5.2).

### 5.2.1　First Stage Identification – IV Estimation

In TSLS, the most important concept is the IV. IV is introduced into the identification procedure to solve the problem of inconsistent OLS identification results caused by omitted variables or the correlation between explanatory variables and error terms.

In the structural equation (5.1), we assume that an important exogenous variable, that is, dew-point temperature $T_{dew}(t)$, other than $T_{dry}(t)$ is omitted. An IV estimation is needed to compensate for the inconsistent estimation results caused by this omission. We choose $T_{dew}(t)$ to be the IV for $T_{dry}(t)$. Thus, $T_{dew}(t)$ should satisfy

$$\text{Cov}[T_{dew}(t), T_{dry}(t)] \neq 0,$$
$$\text{Cov}[T_{dew}(t), \epsilon(t)] = 0. \tag{5.3}$$

Then a simple IV estimation model can be constructed as

$$T_{dry}(t) = \gamma_0 + \gamma_1 T_{dew}(t) + \upsilon(t), \tag{5.4}$$

where $\gamma_0$ and $\gamma_1$ are the estimation parameters in this model; $v(t)$ is the error term, which satisfies

$$E[v(t)] = 0,$$

$$Cov[v(t), T_{dew}(t)] = 0.$$

According to (5.4), the IV estimation for $T_{dry}(t)$ can be denoted as

$$\hat{T}_{dry}(t) = \hat{\gamma}_0 + \hat{\gamma}_1 T_{dew}(t), \tag{5.5}$$

where $\hat{\gamma}_0$ and $\hat{\gamma}_1$ are the estimated values of $\gamma_0$ and $\gamma_1$, respectively.

Let $\{T_{dry}^o(i), T_{dew}^o(i)\}, \forall i = 1, \ldots, n_d$ denote $n_d$ observation pointsof dry-bulb temperature and dew-point temperature, $\hat{\gamma}_0$ and $\hat{\gamma}_1$ can be estimated using OLS as

$$\hat{\gamma}_1 = \frac{Cov[T_{dry}^o, T_{dew}^o]}{Var[T_{dew}^o]}, \tag{5.6}$$

and

$$\hat{\gamma}_0 = \overline{T}_{dry}^o - \hat{\gamma}_1 \overline{T}_{dew}^o. \tag{5.7}$$

The IV estimation calculated from (5.5) will then be used in the second stage of model identification.

## 5.2.2  Second Stage Identification – TSRLS

Inspired by the idea in [17, 19], we adopt the TSRLS to accomplish the second stage of identification. By using (5.5), all historical IV estimations $\hat{T}_{dry}(t)$ can be calculated according to $T_{dew}(t)$. Then the structural equation (5.1) becomes

$$\mathfrak{A}(z)y_h(t) = \alpha_0 + \mathfrak{B}(z)\hat{T}_{dry}(t) + \mathfrak{C}(z)\epsilon(t). \tag{5.8}$$

Define the parameter vectors,

$$\vartheta_\alpha \triangleq [\alpha_1 \; \alpha_2 \; \cdots \; \alpha_{n_\alpha}]^\mathsf{T} \in \mathbb{R}^{n_\alpha},$$

$$\vartheta_\beta \triangleq [1 \; \beta_1 \; \beta_2 \; \cdots \; \beta_{n_\beta}]^\mathsf{T} \in \mathbb{R}^{n_\beta+1}, \tag{5.9}$$

$$\vartheta_\pi \triangleq [\pi_1 \; \pi_2 \; \cdots \; \pi_{n_\pi}]^\mathsf{T} \in \mathbb{R}^{n_\pi},$$

and information vectors,

$$\varphi_\alpha(t) \triangleq [-y_h(t-1) \; \cdots \; -y_h(t-n_\alpha)]^\mathsf{T} \in \mathbb{R}^{n_\alpha},$$

$$\varphi_\beta(t) \triangleq [\hat{T}_{dry}(t) \; \hat{T}_{dry}(t-1) \; \cdots \; \hat{T}_{dry}(t-n_\beta)]^\mathsf{T} \in \mathbb{R}^{n_\beta+1}, \tag{5.10}$$

$$\varphi_\pi(t) \triangleq [\epsilon(t-1) \; \cdots \; \epsilon(t-n_\pi)]^\mathsf{T} \in \mathbb{R}^{n_\pi}.$$

From (5.8), we have

$$
\begin{aligned}
y_h(t) &= [1 - \mathfrak{A}(z)]y_h(t) + \mathfrak{B}(z)\hat{T}_{dry}(t) + \mathfrak{C}(z)\epsilon(t) \\
&= \varphi_\alpha(t)^\mathsf{T}\vartheta_\alpha + \varphi_\beta(t)^\mathsf{T}\vartheta_\beta + \varphi_\pi(t)^\mathsf{T}\vartheta_\pi + \epsilon(t) \\
&= \varphi^\mathsf{T}(t)\vartheta + \epsilon(t),
\end{aligned}
\tag{5.11}
$$

where

$$
\begin{aligned}
\varphi(t) &= [\varphi_\alpha(t)^\mathsf{T} \ \varphi_\beta(t)^\mathsf{T} \ \varphi_\pi(t)^\mathsf{T}]^\mathsf{T} \in \mathbb{R}^n, \\
\vartheta &= [\vartheta_\alpha^\mathsf{T} \ \vartheta_\beta^\mathsf{T} \ \vartheta_\pi^\mathsf{T}]^\mathsf{T} \in \mathbb{R}^n, n = n_\alpha + n_\beta + n_\pi + 1.
\end{aligned}
\tag{5.12}
$$

In order to reduce the size of the problem, the estimation system in (5.11) can be decoupled into two subsystems. Then the auxiliary model identification idea [19] can be used to identify the parameter vectors. The first subsystem is defined to be a classic AR model which can be represented as

$$
\begin{aligned}
y_{h1}(t) &\triangleq y_h(t) - \varphi_{\beta\pi}(t)^\mathsf{T}\vartheta_{\beta\pi} \\
&= \varphi_\alpha^\mathsf{T}\vartheta_\alpha + \epsilon(t),
\end{aligned}
\tag{5.13}
$$

where $\varphi_{\beta\pi} = [\varphi_\beta^\mathsf{T} \ \varphi_\pi^\mathsf{T}]^\mathsf{T}$ and $\vartheta_{\beta\pi} = [\vartheta_\beta^\mathsf{T} \ \vartheta_\pi^\mathsf{T}]^\mathsf{T}$. The second subsystem can be treated as an MA model, and which can be represented as

$$
\begin{aligned}
y_{h2}(t) &\triangleq y_h(t) - \varphi_\alpha(t)^\mathsf{T}\vartheta_\alpha \\
&= \varphi_{\beta\pi}(t)^\mathsf{T}\vartheta_{\beta\pi} + \epsilon(t).
\end{aligned}
\tag{5.14}
$$

In order to use least squares (LS) to identify parameters, two LS cost functions are defined to be

$$
\begin{aligned}
J_{h1}(\vartheta_\alpha) &= \sum_{j=\xi}^{t} \left[y_{h1}(j) - \varphi_\alpha(j)^\mathsf{T}\vartheta_\alpha\right]^2, \\
J_{h2}(\vartheta_{\beta\pi}) &= \sum_{j=\xi}^{t} \left[y_{h2}(j) - \varphi_{\beta\pi}(j)^\mathsf{T}\vartheta_{\beta\pi}\right]^2,
\end{aligned}
\tag{5.15}
$$

where $\xi \geq \max(n_\alpha, n_\beta, n_\pi)$. In the next step, in order to obtain the LS for (5.8), we only need to minimize $J_{h1}(\vartheta_\alpha)$ and $J_{h2}(\vartheta_{\beta\pi})$, respectively. Taking the first order derivatives of $J_{h1}(\vartheta_\alpha)$ and $J_{h2}(\vartheta_{\beta\pi})$ with respect to $\vartheta_\alpha$ and $\vartheta_{\beta\pi}$ and let them be 0, we have

$$
\begin{aligned}
\frac{\mathrm{d}J_{h1}(\vartheta_\alpha)}{\mathrm{d}\vartheta_\alpha} &= -2\varphi_\alpha(j)\sum_{j=\xi}^{t}[y_{h1}(j) - \varphi_\alpha(j)^\mathsf{T}\vartheta_\alpha] = 0, \\
\frac{\mathrm{d}J_{h2}(\vartheta_{\beta\pi})}{\mathrm{d}\vartheta_{\beta\pi}} &= -2\varphi_{\beta\pi}(j)\sum_{j=\xi}^{t}[y_{h2}(j) - \varphi_{\beta\pi}(j)^\mathsf{T}\vartheta_{\beta\pi}] = 0.
\end{aligned}
\tag{5.16}
$$

Let $\hat{\vartheta}(j) = [\hat{\vartheta}_\alpha(j)^\mathsf{T} \ \hat{\vartheta}_{\beta\pi}(j)^\mathsf{T}]^\mathsf{T} \in \mathbb{R}^n$ be the estimation of $\vartheta$ at time $j$. Then by following similar lines as in [17] and combining (5.13) and (5.14), we have the recursive LS procedure

$$\hat{\vartheta}_\alpha(J) = \hat{\vartheta}_\alpha(J-1) + \mathfrak{L}_\alpha(J)[y_h(J) - \varphi_{\beta\pi}^{\mathsf{T}}(J)\vartheta_{\beta\pi} - \varphi_\alpha^{\mathsf{T}}\hat{\vartheta}_\alpha(J-1)], \tag{5.17}$$

$$\mathfrak{L}_\alpha(J) = \mathfrak{P}_\alpha(J-1)\varphi_\alpha(J)[1 + \varphi_\alpha^{\mathsf{T}}\mathfrak{P}_\alpha(J-1)\varphi_\alpha(J)]^{-1}, \tag{5.18}$$

$$\mathfrak{P}_\alpha(J) = [I - \mathfrak{L}_\alpha(J)\varphi_\alpha(J)^{\mathsf{T}}]\mathfrak{P}_\alpha(J-1), \tag{5.19}$$

$$\mathfrak{P}_\alpha(\xi) = p_\xi I, \tag{5.20}$$

$$\hat{\vartheta}_{\beta\pi}(J) = \hat{\vartheta}_{\beta\pi}(J-1) + \mathfrak{L}_{\beta\pi}(J)[y_h(J) - \varphi_\alpha(J)^{\mathsf{T}}\vartheta_\alpha - \varphi_{\beta\pi}^{\mathsf{T}}\hat{\vartheta}_{\beta\pi}(J-1)], \tag{5.21}$$

$$\mathfrak{L}_{\beta\pi}(J) = \mathfrak{P}_{\beta\pi}(J-1)\varphi_{\beta\pi}(J)[1 + \varphi_{\beta\pi}^{\mathsf{T}}\mathfrak{P}_{\beta\pi}(J-1)\varphi_{\beta\pi}(J)]^{-1}, \tag{5.22}$$

$$\mathfrak{P}_{\beta\pi}(J) = [I - \mathfrak{L}_{\beta\pi}(J)\varphi_{\beta\pi}(J)^{\mathsf{T}}]\mathfrak{P}_{\beta\pi}(J-1), \tag{5.23}$$

$$\mathfrak{P}_{\beta\pi}(\xi) = p_\xi I. \tag{5.24}$$

Because $\vartheta_\alpha$ and $\vartheta_{\beta\pi}$ in (5.17) and (5.21) cannot be obtained in reality, we use the previous estimations for $\vartheta_\alpha$ and $\vartheta_{\beta\pi}$ at time $J$ as

$$\vartheta_{\beta\pi} = \hat{\vartheta}_{\beta\pi}(J-1),$$
$$\vartheta_\alpha = \hat{\vartheta}_\alpha(J-1). \tag{5.25}$$

This results in

$$\hat{\vartheta}_\alpha(J) = \hat{\vartheta}_\alpha(J-1) + \mathfrak{L}_\alpha(J)[y_h(J) - \varphi(J)^{\mathsf{T}}\hat{\vartheta}(J-1)], \tag{5.26}$$

$$\hat{\vartheta}_{\beta\pi}(J) = \hat{\vartheta}_{\beta\pi}(J-1) + \mathfrak{L}_{\beta\pi}(J)[y_h(J) - \varphi(J)^{\mathsf{T}}\hat{\vartheta}(J-1)]. \tag{5.27}$$

By replacing (5.17) and (5.21) with (5.26) and (5.27), we can readily obtain the TSRLS algorithm for the second stage in TSLS. The flow chart of the whole identification procedure is shown in Figure 5.1. This procedure will not end until it reaches the most recent available data.

After performing the two-stage estimation, the resulting model can be represented as

$$y_h(t) = \hat{\alpha}_0 - \sum_{i=1}^{n_\alpha} \hat{\alpha}_i z^{-i} y_h(t) + \sum_{j=0}^{n_\beta} \hat{\beta}_j z^{-j} T_{dry}(t) + \sum_{k=1}^{n_\pi} \hat{\pi}_k z^{-k} \epsilon(t) \tag{5.28}$$

Note that, in a simple AR model, the constant term $\alpha_0$ in (5.8) can represent a bias for the model; however, in a complex model like (5.8), since a lot of lags may be included, this bias term $\alpha_0$ is not significant. Thus, $\alpha_0$ is set to be 0 when we identify the model; but, we still put $\alpha_0$ and $\hat{\alpha}_0$ in (5.8) and (5.28) to keep the completeness.

## 5.2.3   Load Forecasting

The identification method for the forecasting model mentioned above is to deal with the heating load $y_h(t)$. For cooling and electrical loads $y_c(t)$ and $y_e(t)$, respectively, the forecasting models and identification procedures are the same. The forecasted loads,

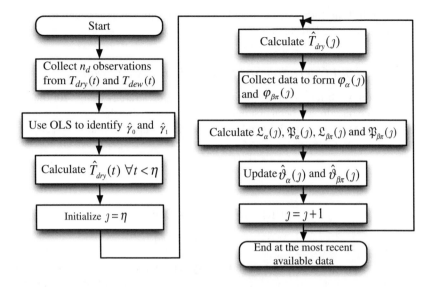

**Figure 5.1** OLS-TSRLS algorithm flowchart

that is, $\hat{y}_h(t)$, $\hat{y}_c(t)$, and $\hat{y}_e(t)$, can be readily obtained by using the identified model and historical data. At every step, the actual value is obtained, the forecasted value will be immediately replaced by the actual one, which can be used in the next step of forecast. Note that, on the one hand, the coefficients of the forecasting model cannot be guaranteed to be all positive. Hence, even though all historical data are positive, the forecasted value still has the possibility to be negative. On the other hand, since the error term $\epsilon(t)$, which can either be positive or negative, is involved in the forecasting model, if the forecasted value is sufficiently small or equal to 0, this error term may drive the forecasting result to go below 0. It is worth mentioning that these negative forecasting values only exist when the forecasted value is small or equal to 0; when the forecasted value is large enough, the error term or negative coefficients cannot affect the result a lot. Thus, one simple way to deal with this phenomenon is to regulate these negative forecasting results to 0.

## 5.3 Operation Strategy Design

After obtaining the forecasted loads in time interval $[t, t + 1]$, we can start using the operation strategy to meet the demands. However, in most cases, the forecasted data are not the same as the actual data. Hence, after fulfilling the forecasted demands, a post-strategy has to be used to coordinate the energy flow in the system or to appropriately purchase additional primary energy to compensate for the gap between forecasted and actual loads. Thus, the operation strategy design for the CCHP system using forecasted loads should be divided into two steps: (1) design a strategy to optimally fulfill the forecasted demands; and (2) design a post-strategy to deal with the inaccuracy of the forecasting.

### 5.3.1 Optimal Operation Strategy for Forecasted Load

In this section, we briefly review the optimal operation strategy design for the CCHP system, which has already been discussed in Chapter 4. Using the energy hub concept, the CCHP system in Figure 3.1 can be modeled in an input–output form [20] as

$$\hat{\mathcal{V}}_o(t) = \mathcal{H}\mathcal{V}_i(t), \tag{5.29}$$

where

$$\mathcal{V}_i(t) = [F_m(t) \, E_{grid}(t) \, 0 \, 0]^{\mathsf{T}}, \tag{5.30}$$

$$\hat{\mathcal{V}}_o(t) = [0 \, \hat{y}_e(t) \, \hat{y}_c(t) \, \hat{y}_h(t)]^{\mathsf{T}}, \tag{5.31}$$

represent the system input vector and the system output (forecasted load) vector during the time interval $[t, t+1]$, respectively; and the system transition matrix $\mathcal{H}$ can be found in (4.2). In (5.30), $F_m(t)$ and $E_{grid}(t)$ are the total fuel and electricity purchased during $[t, t+1]$, respectively. Herein, $\eta_{pgu}$, $\eta_b$, $\eta_{hrs}$, and $\eta_h$ denote the efficiencies of the PGU, auxiliary boiler, heating recovery system, and heating unit, respectively; $COP_{ec}$ and $COP_{ac}$ represent coefficients of performance of the electric chiller and absorption chiller, respectively. In addition, $\alpha_{pgu}$ and $\alpha_b$ are the dispatch factors denoting the fuel fraction flows into the PGU and boiler, respectively, and $\alpha_{pgu} + \alpha_b = 1$; $\alpha_{user}$ and $\alpha_{ec}$ are the dispatch factors denoting the electricity fraction flows to the terminal users and electric chiller, respectively, and $\alpha_{user} + \alpha_{ec} = 1$; $\alpha_{ac}$ and $\alpha_h$ are the dispatch factors denoting the heat fraction flows to the absorption chiller and heating unit, respectviely, and $\alpha_{ac} + \alpha_h = 1$.

The optimal operation strategy for the forecasted load $\hat{\mathcal{V}}_o(t)$ can be found in Section 4.3.

The optimization can be solved either by using SQP or the algorithm proposed in the appendix at the end of this chapter. Once the optimization is completed, optimal energy input for the forecasted loads during the time interval $[t, t+1]$ can be obtained as

$$\mathcal{V}_i^*(t) = [F_m^*(t) \, E_{grid}^*(t) \, 0 \, 0]^{\mathsf{T}}. \tag{5.32}$$

### 5.3.2 Post-strategy Design

If the actual loads during the time interval $[t, t+1]$ are defined as

$$\mathcal{V}_o(t) = [0 \, y_e(t) \, y_c(t) \, y_h(t)]^{\mathsf{T}}, \tag{5.33}$$

there should exist discrepancies between $\hat{\mathcal{V}}_o(t)$ and $\mathcal{V}_o(t)$. Thus, we either need to purchase additional primary energy, or need to sell back or simply discard the excess energy. The general idea in this section is to use the surplus of one type of energy, for example, electricity, to compensate for the gap of another type of energy, for example, cooling, through coordinating the power flow within the system, for example, by using the electric chiller. Since there exist three loads to be discussed, the relationship between the forecasted loads and actual ones can be divided into the following eight cases.

1. $\hat{y}_e(t) \geq y_e(t)$, $\hat{y}_h(t) \geq y_h(t)$, $\hat{y}_c(t) \geq y_c(t)$
2. $\hat{y}_e(t) \geq y_e(t)$, $\hat{y}_h(t) \geq y_h(t)$, $\hat{y}_c(t) < y_c(t)$
3. $\hat{y}_e(t) \geq y_e(t)$, $\hat{y}_h(t) < y_h(t)$, $\hat{y}_c(t) \geq y_c(t)$
4. $\hat{y}_e(t) \geq y_e(t)$, $\hat{y}_h(t) < y_h(t)$, $\hat{y}_c(t) < y_c(t)$
5. $\hat{y}_e(t) < y_e(t)$, $\hat{y}_h(t) \geq y_h(t)$, $\hat{y}_c(t) \geq y_c(t)$
6. $\hat{y}_e(t) < y_e(t)$, $\hat{y}_h(t) \geq y_h(t)$, $\hat{y}_c(t) < y_c(t)$
7. $\hat{y}_e(t) < y_e(t)$, $\hat{y}_h(t) < y_h(t)$, $\hat{y}_c(t) \geq y_c(t)$
8. $\hat{y}_e(t) < y_e(t)$, $\hat{y}_h(t) < y_h(t)$, $\hat{y}_c(t) < y_c(t)$

### 5.3.2.1 $\hat{y}_e(t) \geq y_e(t), \hat{y}_h(t) \geq y_h(t), \hat{y}_c(t) \geq y_c(t)$

In this case, all of the three forecasted loads exceed or are equal to the corresponding actual loads. Since we assume no heat storage system in the system configuration, the excess energy cannot be stored for the next hour's use. Thus, except for selling back the excess electricity, no action needs to be taken under this situation, and this part of the energy will definitely be wasted. This is the most inefficient case.

### 5.3.2.2 $\hat{y}_e(t) \geq y_e(t), \hat{y}_h(t) \geq y_h(t), \hat{y}_c(t) < y_c(t)$

In this case, it is inefficient to simply purchase electricity or fuel to drive the electric chiller or absorption chiller without using the excess heating energy. Thus, we may consider making use of the excess heating energy and/or electricity to fill up the cooling demand gap. The reason for the thermal energy being transferable is because we can always arbitrarily dispatch the thermal flow between the heating unit and the absorption chiller or dispatch the electricity between the users and electric chiller. In this case, as soon as the heating load or electrical load is fully covered, while the cooling is still being provided, we can dispatch the excess energy to the absorption chiller and electric chiller, respectively, to provide additional cooling. In order to achieve this, it is necessary to check if the excess electricity and heating can fill up the cooling demand gap, which implies

$$[\hat{y}_e(t) - y_e(t)]COP_{ec} + [\hat{y}_h(t) - y_h(t)]\frac{COP_{ac}}{\eta_h} \geq y_c(t) - \hat{y}_c(t). \qquad (5.34)$$

If (5.34) holds, neither additional electricity nor fuel needs to be purchased but the remaining excess electricity after the compensating should be sold back; if not, the rest of the cooling that needs to be provided can be represented by

$$\tilde{y}_c(t) = y_c(t) - \hat{y}_c(t) - [\hat{y}_e(t) - y_e(t)]COP_{ec} - [\hat{y}_h(t) - y_h(t)]\frac{COP_{ac}}{\eta_h}. \qquad (5.35)$$

Instead of running the PGU to provide this part of the energy, we choose to use other direct ways to do so, that is, running the auxiliary boiler to drive the absorption chiller or directly purchasing electricity to run the electric chiller. The decision making between the electric chiller and absorption chiller should depend on the actual cost function values, however, in practice, it is impossible to know how much more cooling needs to be provided. Thus, chiller selection according to the cost function values cannot be achieved. Considering this and the efficiency, the electric chiller will be solely used to meet $\tilde{y}_c(t)$, which implies that the additional electricity that needs to be purchased during the time interval $[t, t + 1]$ is

$$\check{E}_{grid}(t) = \frac{\tilde{y}_c(t)}{COP_{ec}}. \tag{5.36}$$

### 5.3.2.3    $\hat{y}_e(t) \geq y_e(t), \hat{y}_h(t) < y_h(t), \hat{y}_c(t) \geq y_c(t)$

In this case, we can use a similar idea to the previous case, which means transferring the excess energy to the heating unit. While the excess electricity provided cannot help in heating, the excess thermal energy from the absorption chiller can be transferred to the heating unit. To do so, it is necessary to check if this part of the thermal energy can fully cover the heating shortage, which implies

$$\frac{[\hat{y}_c(t) - y_c(t)](1 - \zeta)}{COP_{ac}}\eta_h \geq y_h(t) - \hat{y}_h(t), \tag{5.37}$$

where $\zeta$ is the electric cooling to cool load ratio. If (5.37) holds, there is no need to purchase any primary energy; if not, the rest of the heating needed can be represented as

$$\tilde{y}_h(t) = y_h(t) - \hat{y}_h(t) - \frac{[(\hat{y}_c(t) - y_c(t)](1 - \zeta)}{COP_{ac}}\eta_h. \tag{5.38}$$

As a result, additional fuel needs to be purchased to run the absorption chiller to meet $\tilde{y}_h(t)$ can be calculated as

$$\check{F}_m(t) = \frac{\tilde{y}_h(t)}{\eta_h\eta_b}. \tag{5.39}$$

The excess electricity should be sold back to the grid.

### 5.3.2.4    $\hat{y}_e(t) \geq y_e(t), \hat{y}_h(t) < y_h(t), \hat{y}_c(t) < y_c(t)$

In this case, the excess electricity can be used to generate part of the cooling shortage, however, the heating gap can only be filled up by purchasing fuel to run the auxiliary boiler. First, the condition that is used to verify that the excess electricity can fully compensate for the demand gap should be checked. This implies

$$[\hat{y}_e(t) - y_e(t)]COP_{ec} \geq y_c(t) - \hat{y}_c(t). \tag{5.40}$$

If (5.40) holds, the cooling gap can be filled up without purchasing any primary energy, but the remaining excess electricity after the compensating should be sold back; if not, the rest of the cooling

$$\tilde{y}_c(t) = y_c(t) - \hat{y}_c(t) - [\hat{y}_e(t) - y_e(t)]COP_{ec}, \tag{5.41}$$

should be provided by the electric chiller as stated in Section 5.3.2.2. Doing so will cause additional electricity purchasing from the local grid. The additional electricity $\check{E}_{grid}(t)$ can be calculated using (5.36). The heating demand gap will be solely provided by the auxiliary boiler. The additional fuel to be purchased can be represented as

$$\check{F}_m(t) = \frac{y_h(t) - \hat{y}_h(t)}{\eta_h \eta_b}. \tag{5.42}$$

### 5.3.2.5   $\hat{y}_e(t) < y_e(t), \hat{y}_h(t) \geq y_h(t), \hat{y}_c(t) \geq y_c(t)$

In this case, since neither the excess heating nor the excess cooling energy can be transferred to provide electricity, the electrical demand gap should be solely filled up by directly purchasing from the local grid. This implies

$$\check{E}_{grid}(t) = y_e(t) - \hat{y}_e(t). \tag{5.43}$$

### 5.3.2.6   $\hat{y}_e(t) < y_e(t), \hat{y}_h(t) \geq y_h(t), \hat{y}_c(t) < y_c(t)$

In this case, as stated before, the excess heating energy can be transferred to compensate for part of or all of the cooling demand shortage, however, the electricity shortage should be directly purchased from the local grid. First, it is necessary to check if the excess heating energy can solely provide the cooling demand shortage, which indicates

$$[\hat{y}_h(t) - y_h(t)]\frac{COP_{ac}}{\eta_h} \geq y_c(t) - \hat{y}_c(t). \tag{5.44}$$

If (5.44) holds, there is no need to purchase any primary energy to provide cooling; if not, the rest of the cooling demand can be represented as

$$\tilde{y}_c(t) = y_c(t) - \hat{y}_c(t) - [\hat{y}_h(t) - y_h(t)]\frac{COP_{ac}}{\eta_h}. \tag{5.45}$$

Same as the previous case, the electrical shortage will be directly purchased from the local grid. Combining the electricity needed by the electric chiller and the direct electricity shortage, the additional electricity that needs to be purchased from the local grid can be calculated as

$$\check{E}_{grid}(t) = y_e(t) - \hat{y}_e(t) + \frac{\tilde{y}_c(t)}{COP_{ec}}. \tag{5.46}$$

**5.3.2.7**   $\hat{y}_e(t) < y_e(t), \hat{y}_h(t) < y_h(t), \hat{y}_c(t) \geq y_c(t)$

In this case, part of or all of the heating demand gap can be filled up by the excess cooling from the absorption chiller, however, the electricity shortage should be directly purchased from the local grid. Thus, the condition to verify that the excess cooling from the absorption chiller part can do this job should be checked as in (5.37). If this condition holds, no additional fuel needs to be purchased; if not, the additional purchased fuel can be represented as (5.39). Moreover, the electricity shortage that will be directly purchased can be calculated by using (5.43).

**5.3.2.8**   $\hat{y}_e(t) < y_e(t), \hat{y}_h(t) < y_h(t), \hat{y}_c(t) < y_c(t)$

In this case, none of the forecasted electrical, heating and cooling loads is enough. Thus, considering the previous seven cases, the heating shortage will be provided by the auxiliary boiler; the cooling shortage and electricity shortage will be provided by the electric chiller and purchasing from the local grid directly, respectively. The fuel to be purchased can be calculated using (5.42); electricity needs to be purchased from the local grid can be calculated as

$$\check{E}_{grid}(t) = y_e(t) - \hat{y}_e(t) + \frac{y_c(t) - \hat{y}_c(t)}{COP_{ec}}. \tag{5.47}$$

## 5.4   Case Study

The above two sections present the detailed model identification, forecasting and post-strategy design procedure. To validate the feasibility and effectiveness, we conduct a case study to verify the accuracy of the identified forecasting model and the efficiency of the post-strategy.

### 5.4.1   Hypothetical Building Configuration

The hypothetical building configuration can be referred to Section 3.4.1.

### 5.4.2   Weather and Load Data

The weather data (from year 2002 to year 2011) of Victoria is collected from the National Renewable Energy Laboratory, the US DOE, and the Environment Canada. The meteorological station is located in YYJ (Victoria International Airport). For each hour, the weather data consist of dry-bulb temperature (°C), dew-point temperature (°C), wind direction (°), wind speed (m/s), altimeter (hundreds of pascals), and visibility (m). Considering the continuity of the data observations, we choose the data from year 2002 to year 2008 as the sample, and choose the data of year 2011 to forecast.

First, EnergyPlus [21] is adopted to use the weather data to simulate the electrical, heating and cooling loads for this hypothetical building.

Having the historical loads, in the second stage, EViews [22] is chosen to analyze the load data. From year 2002 to year 2008, with leap years considered, 61 368

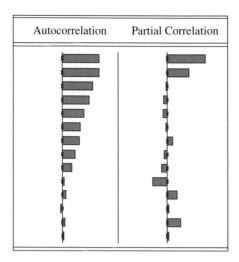

**Figure 5.2**    Heating loads correlogram

samples in total for each load are analyzed. The correlograms, which can be used to determine the number of lags in AR terms, of heating, electric and cooling loads produced by EViews are shown in Figures 5.2–5.4, respectively. Each row of the correlogram represents the character of a specific lag, that is, lag 1, lag 2, ..., lag 14 from the top to the bottom. In the autocorrelation part of the correlogram, for example, if the shadow bar of lag 1 exceeds the confidence interval (narrow area close to the middle in Figures 5.2–5.4), the hypothesis is rejected, which indicates the current data are correlated with their 1-lag data; in the partial correlation part, for example, if the shadow bar of lag 10 data exceeds the confidence interval, the hypothesis is rejected, which implies that 10 lags jointly correlate with the current data.

Usually, we can determine the number of lags in AR terms $(n_\alpha)$ by observing after which lag the hypothesis can be accepted. However, from these three figures, we can observe significant autocorrelations among these samples even at lag 14. If the correlogram of more data is presented, for example 999 lags, the autocorrelation term shows a significant periodic autocorrelation, which indicates that strong and weak autocorrelation lags alternatively exist as the number of lags increases. Theoretically, it is recommended to include all lags with strong autocorrelation in the forecasting model. However, since too many lags will dramatically increase the computational load, it is not practical to include all of them. To obtain the best $n_\alpha$, we conducted a parametric analysis and found that, when $n_\alpha$ is around 500, all criteria, that is, mean square prediction error (MSPE), bias, mean absolute error (MAE), and mean absolute percentage error (MAPE), reach the lowest value. Since every lag stands for 1 h, 504 lags are just 21 days (3 weeks), which is more reasonable than 500 lags. Thus, $n_\alpha = 504$ was selected in this chapter for the later case study. From the partial correlation part of the correlograms, we can see all of the three data sets have relatively small partial correlations after 3 lags. However, since it is hard to detect the number of MA

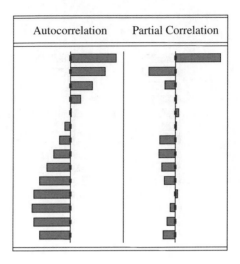

**Figure 5.3**  Electrical loads correlogram

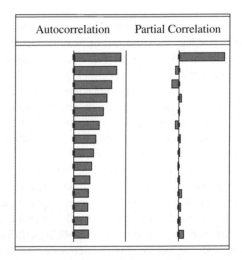

**Figure 5.4**  Cooling loads correlogram

lags from the partial correlation part due to the effect of AR terms, by trial and error, we chose 2 lags for the MA term in the forecasting model ($n_\pi = 2$). The exogenous input for the forecasting model is chosen to be the dry-bulb temperature $T_{dry}$, and the IV for $T_{dry}$ is chosen to be the dew-point temperature $T_{dew}$. In this chapter, $n_\beta$ is set to be 0, which implies that only the current dry-bulb temperature is being used to forecast the loads. $n_\beta$ is set in this way because (1) the current load depends more on the current dry-bulb temperature than any previous time dry-bulb temperature, and (2) considering the large $n_\alpha$, setting $n_\beta = 0$ can help to reduce the computational load.

Adopting the parameter identification approach in Section 5.2, we have three sets of $\alpha_i$ $(i = 1, 2, \ldots, n_\alpha)$, $\beta_j$ $(j = 0, 1, 2, \ldots, n_\beta)$ and $\pi_k$ $(k = 1, 2, \ldots, n_\pi)$ identified for heating, electrical and cooling loads model. This identification was conducted in MATLAB with Intel i7 2.8 GHz and 16 GB memory without parallel computing. The time consumed by the proposed OLS-TSRLS algorithm to identify the forecasting model is 197 s, while it takes 401 s using conventional TSLS. The advantage of the proposed OLS-TSRLS algorithm is because of the reduced time complexity, which is $O(n_\alpha^2)$, while the time complexity of the conventional TSLS is $O(n_\alpha^3)$. Having the identified model and the data from year 2002 to year 2008 as the sample, the loads in year 2011 can be forecasted. The benchmark actual load can be simulated by assuming that all the loads are already known. The methodology is combining the 7 years' historical data as a time series and forecasting the load using the identified model. The forecasted heating load and error, forecasted electrical load and error, and forecasted cooling load and error are shown in Figures 5.5–5.10. Note that, since the forecasted electrical loads are close to the actual electrical loads, the plot may convey less information if data from all 8640 h are included. Thus, we only pick up a random day to show the electrical loads as in Figure 5.7.

From Figure 5.5, it can be observed that the variation trend of the actual and forecasted heating loads are almost the same, which means, though errors exist, peak and bottom points of the two loads stay in the same time interval. However, when there exists a sudden change in the actual loads, such as the small spike between 5530 and 5540 in Figure 5.5, the forecasted data cannot catch this change. As a result, a 1-h lag will be generated in the forecasted loads. Except for this situation, most of the forecasted data can well catch the variations of the actual loads. In Figure 5.6, most of the errors stay in the range $[-10, 10]$ kWh, while few can reach

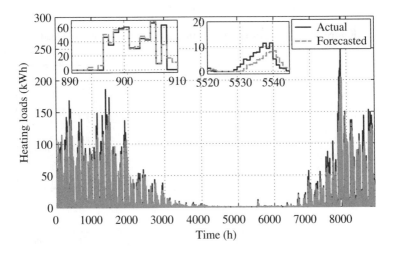

**Figure 5.5**  Comparison between forecasted and actual heating loads

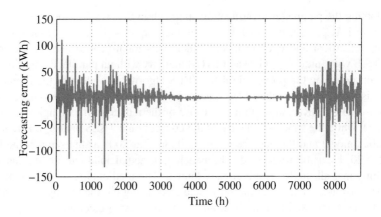

**Figure 5.6**   Error between forecasted and actual heating loads

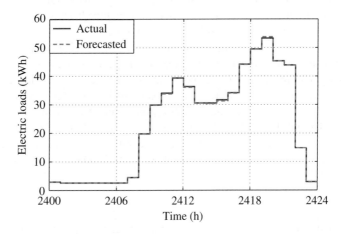

**Figure 5.7**   Comparison between forecasted and actual electric loads

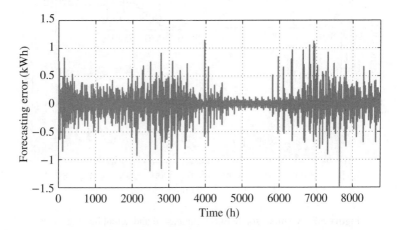

**Figure 5.8**   Error between forecasted and actual electrical loads

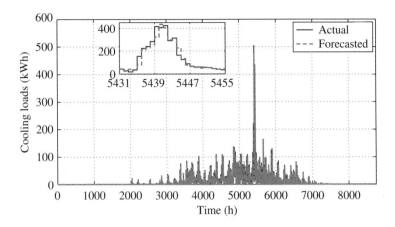

**Figure 5.9**   Comparison between forecasted and actual cooling loads

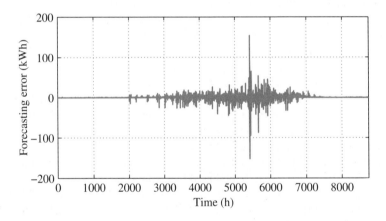

**Figure 5.10**   Error between forecasted and actual cooling loads

the maximum value, which is around $\pm 120$ kWh. This large error is caused by the sudden unusual change in the actual heating load, which is normally due to a sudden connection of a large number of heating devices or a high power heating device. This scenario is quite normal in winter. Though the error in this case becomes large, the forecasting gap can be well compensated for by the proposed post-strategy so that the energy consumption, total cost, and $CO_2$ emissions are only slightly increased or even decreased when compared with the CCHP system using actual loads, which shall be shown later. For the electrical loads, the result is much better. The forecasted electrical load can match the actual electrical load quite well. From Figure 5.8, we can see the maximum error is no more than $\pm 1.5$ kWh. For the cooling **load**, from Figure 5.10, only one large error happens in the middle of summer, which is also caused by the sudden change of actual cooling load. Except for this spike, almost all of the other cooling forecasting errors are within the range $[-10, +10]$ kWh.

**Table 5.1** Average normed error of different models using six sets of $n_\alpha$, $n_\beta$, and $n_\pi$

| $(n_\alpha, n_\beta, n_\pi)$ | OLS-TSRLS | TSLS | ARMAX (no IV) |
|---|---|---|---|
| $(504, 0, 2)$ | 18.0% | 24.3% | 35.8% |
| $(24, 0, 2)$ | 24.2% | 33.2% | 41.4% |
| $(672, 0, 2)$ | 20.7% | 25.0% | 36.7% |
| $(504, 2, 2)$ | 19.4% | 24.8% | 36.3% |
| $(504, 0, 0)$ | 21.2% | 26.3% | 35.8% |
| $(504, 0, 10)$ | 23.3% | 31.5% | 35.8% |

Using the parameter set $(n_\alpha, n_\beta, n_\pi) = (504, 0, 2)$, the average relative forecasting error of the three loads by using the forecasting model identified by the proposed OLS-TSRLS algorithm is 18%, which is 25% less than the results obtained from the forecasting model identified by the conventional TSLS. This is because, when using the recursive LS, the load evolution trend caused by the weather profile change can be caught as the recursive process evolves. However, standard TSLS just takes the whole set of data to identify without considering the load evolution trend. Note that, one of the problems involved in the time series estimation is the selection of lags. As stated in Section 5.4.2, the lag selection in this chapter was based on the autocorrelation, partial correlation, and trial and error results. Only the set with the most accurate forecasting results (smallest normed error) is picked to conduct the case study. For the readers' comparison, we have also included several different sets of lags to show the average normed error of all the three loads, which can be found in Table 5.1. In Table 5.1, the forecasting error of ARMAX model with no IV is included to show the significance of involving IV. It can be readily observed that, compared with the ARMAX model without using IV (dew-point temperature), standard TSLS gives better forecasting results; among all three models, ARMAX identified by the proposed OLS-TSRLS gives the most accurate forecasting results.

From the above figures and analyses it can be seen that the model identified by the proposed algorithm can well predict the heating, electrical and cooling loads for the next hour. The load error caused by the inaccurate forecasting can be compensated for by the post-strategy proposed in Section 5.3.2.

### 5.4.3 Test Results

To simulate the PEC, ATC, and CDE, we adopt the CCHP system configuration parameters as in Section 4.4.2. Specifically, we choose $\omega_1$, $\omega_2$, and $\omega_3$ all equal to 1/3 to balance the cost/primary energy savings and GHG emission reductions.

Using the forecasted loads and the operation strategy proposed in Section 5.3, we can readily obtain the optimal energy input to the system and optimal power flow inside the CCHP system. To verify the effectiveness of the proposed post-strategy, we also compare the results of using another strategy, that is, directly purchasing

**Table 5.2**  Performance of different systems using forecasted data obtained from the proposed prediction method

| System | PEC (kWh) | ATC ($) | CDE (g) |
|---|---|---|---|
| SP | 875 853 | 144 713 | 262 693 675 |
| CCHP (Post-strategy) | 797 102 | 50 491 | 226 524 001 |
| CCHP (Direct purchasing) | 829 141 | 51 954 | 235 253 497 |
| CCHP (Actual loads) | 790 088 | 49 322 | 227 007 783 |

**Table 5.3**  Performance of different systems using 1-lag forecasted data

| System | PEC (kWh) | ATC ($) | CDE (g) |
|---|---|---|---|
| SP | 943 931 | 148 092 | 290 282 974 |
| CCHP (Post-strategy) | 857 837 | 51 454 | 248 180 816 |
| CCHP (Direct purchasing) | 933 348 | 55 226 | 267 843 229 |
| CCHP (Actual loads) | 790 088 | 49 322 | 227 007 783 |

**Table 5.4**  Performance of different systems using TSLS forecasted data

| System | PEC (kWh) | ATC ($) | CDE (g) |
|---|---|---|---|
| SP | 901 345 | 147 897 | 279 453 980 |
| CCHP (Post-strategy) | 823 472 | 51 032 | 240 893 129 |
| CCHP (Direct purchasing) | 839 273 | 53 421 | 259 058 347 |
| CCHP (Actual loads) | 790 088 | 49 322 | 227 007 783 |

electricity/fuel without considering rescheduling the power flow as in the proposed post-strategy, to compensate for the electrical/thermal gap. The comparisons among the SP system and CCHP systems using different post-strategies can be found in Table 5.2.

In classic time series forecasting approaches, there also exist other ways to forecast loads, for example, the 1-lag forecasting and ARMAX model (identified by standard TSLS) forecasting. The comparisons among different systems using 1-lag forecasted data and TSLS forecasted data are shown in Table 5.3 and Table 5.4, respectively.

In order to make the comparisons clearer, three column charts in Figures 5.11–5.13 are made according to the data in Tables 5.2–5.4 to compare the PEC, ATC, and CDE, respectively. Note that, for the CCHP system using the actual loads, since no forecasting is needed, we only include one column for each criterion.

From Table 5.2, compared with the SP system (row 1) and CCHP system using direct purchasing as the post-strategy (row 3), the CCHP system using the proposed post-strategy (row 2) has relatively low PEC, ATC, and CDE. Assume that we can obtain the actual loads, the three EC of the CCHP system under the proposed post-strategy are shown in the last row of Table 5.2. Compared with the last row,

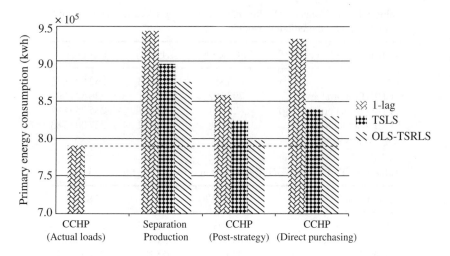

**Figure 5.11**  Comparison of PES

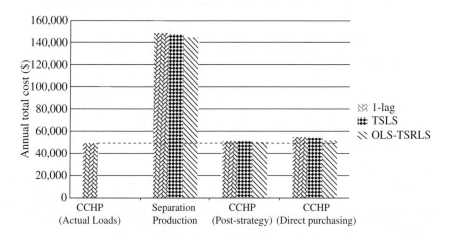

**Figure 5.12**  Comparison of ATC

PEC and ATC of the CCHP system using forecasted loads are slightly increased by 0.9% and 2.4%, respectively; CDE is decreased a little bit by 0.2%. These results show that (1) the forecasted data can well match the actual data with small errors, otherwise, no strategy can guarantee low increase of energy consumption and cost, and (2) the post-strategy does help to save a certain amount of energy.

To compare three different forecasting methods, we can observe the data in Tables 5.3 and 5.4 or read the columns in Figures 5.11–5.13. Compared with the corresponding data in Table 5.2, three criteria in Tables 5.3 and 5.4 are all higher. It is clearly shown in Figures 5.11–5.13 that (1) among three systems, the CCHP system with the proposed post-strategy gives the lowest values of all three criteria, which are almost the same height as the ones of the CCHP system using actual loads; and

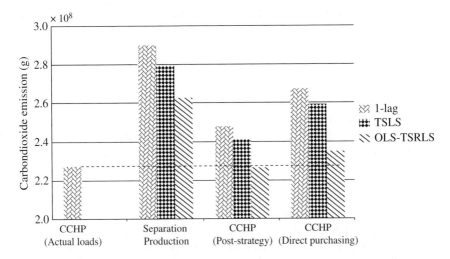

**Figure 5.13**   Comparison of CDE

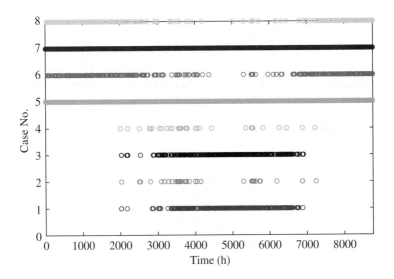

**Figure 5.14**   Case distribution

(2) among three identification approaches, the proposed OLS-TSRLS gives the most accurate forecasting results which can further improve the criteria. Among all the shown columns, the CCHP system using the OLS-TSRLS forecasted loads with the proposed post-strategy gives the proximal performance of the CCHP system using the actual loads.

In the case study of this chapter, within 1 year (8760 h), there are 501, 39, 509, 39, 3233, 591, 3340, and 508 sets of data for eight post-strategy cases, respectively. Their distribution is shown in Figure 5.14, in which the ticks of the $y$-axis represent the case number. It can be readily observed from the distribution that most of the cases fall into Cases 5 and 7.

## 5.5   Summary

In this chapter, we first proposed an ARMAX model, which used current and historical dry-bulb temperature as the exogenous variable, for forecasting the future cooling, heating and electrical loads in a CCHP system. The parameter identification algorithm structure chosen in this chapter was the TSLS, in which the IV for the exogenous variable was selected to be the dew-point temperature. In the first stage of identification, OLS was used to obtain the IV estimation for dry-bulb temperature; in the second stage, TSRLS, which could reduce computational load and could speed up the convergence, was used to identify the parameters in the forecasting model. The loads were predicted using the obtained forecasting model. A post-strategy was designed to compensate for the inaccurate forecasting. The results of the case study showed that forecasted loads obtained from the proposed prediction model could match the actual ones well; and by using the proposed post-strategy, the PEC and ATC of the CCHP system using forecasted loads only increased slightly and CDE even decreased when compared with the case using actual loads. The results of this chapter can also be used in CHP systems.

## Appendix 5.A   Closed-Form Identification Solution of Quadratic ARMAX Model

### 5.A.1   Quadratic ARMA Model

The quadratic ARMA model can be represented as

$$\hat{x}(t) = \mathbf{x}^{\mathrm{T}}(t)\,\mathbf{H}\mathbf{x}(t) + \mathbf{x}^{\mathrm{T}}(t)\mathbf{p} + r, \tag{5.A.1}$$

where

$$\mathbf{x}(t) = \begin{bmatrix} x(t-1) \\ x(t-2) \\ \vdots \\ x(t-n) \end{bmatrix} \in \mathbb{R}^n \tag{5.A.2}$$

is the historical load data, that is, electric, heating and cooling loads, and $\mathbf{H} \in S^n, \mathbf{p} \in \mathbb{R}^n$. $\hat{x}(t)$ is the forecasted heating load at time $t$ and $x(t)$ is the vector of historical load at time $t$. $r$ is a constant. In order to identify the model, that is, to identify $\mathbf{H}$, $\mathbf{p}$, and $r$, the optimization problem is formulated in a least-square way as

$$\min \sum_{t=\zeta}^{T} (\hat{x}(t) - x(t))^2, \tag{5.A.3}$$

where $\zeta \geq n$. Apparently, this problem is an unconstrained convex programming problem. It seems like can be readily solved by using "*cvx*". However, in practice, if the size of vector $x(t)$ or historical data is too large, "*cvx*" cannot handle it. For example, by using "*cvx*", $n = 20$ and $T - \zeta = 1000$ can easily make MATLAB out of memory, while practical identification has $T - \zeta \geq 3 \times 10^4$. In order to tackle this problem, we shall derive a closed-form solution for (5.A.3) to reduce the computational load.

We may use $\hat{x}_t$, $x_t$, and $x_i$ to denote $\hat{x}(t)$, $x(t)$, and $x(t - i)$, respectively. As a result, (5.A.1) becomes

$$\hat{x}_t = \mathbf{x}_t^\mathsf{T} \mathbf{H} \mathbf{x}_t + \mathbf{x}_t^\mathsf{T} \mathbf{p} + r. \tag{5.A.4}$$

Consequently, we can readily have

$$\mathbf{x}_t^\mathsf{T} \mathbf{H} \mathbf{x}_t = [x_1 \, x_2 \, \cdots \, x_n] \begin{bmatrix} h_{11} & * & \cdots & & * \\ h_{21} & h_{22} & * & \cdots & * \\ \vdots & \ddots & \ddots & & \vdots \\ \vdots & & \ddots & \ddots & * \\ h_{n1} & \cdots & \cdots & h_{n(n-1)} & h_{nn} \end{bmatrix} \begin{bmatrix} x_1 \\ x_2 \\ \vdots \\ x_n \end{bmatrix}$$

$$= \left[ \sum_{i=1}^{n} x_i h_{i1} \; \sum_{i=1}^{n} x_i h_{i2} \; \cdots \; \sum_{i=1}^{n} x_i h_{in} \right] \begin{bmatrix} x_1 \\ x_2 \\ \vdots \\ x_n \end{bmatrix}$$

$$= \sum_{i=1}^{n} x_i h_{i1} x_1 + \sum_{i=1}^{n} x_i h_{i2} x_2 + \cdots + \sum_{i=1}^{n} x_i h_{in} x_n \tag{5.A.5}$$

$$= \sum_{i=1}^{n} \sum_{j=1}^{n} h_{ij} x_i x_j$$

$$= \tilde{\mathbf{h}}^\mathsf{T} \tilde{\mathbf{x}}_t,$$

where

$$\tilde{\mathbf{h}} = \begin{bmatrix} h_{11} \\ h_{21} \\ \vdots \\ h_{n1} \\ h_{22} \\ \vdots \\ h_{n2} \\ \vdots \\ h_{nn} \end{bmatrix} \in \mathbb{R}^{n(n+1)/2}, \; \tilde{\mathbf{x}}_t = \begin{bmatrix} x_1^2 \\ 2x_2 x_1 \\ 2x_3 x_1 \\ \vdots \\ 2x_n x_1 \\ x_2^2 \\ 2x_3 x_2 \\ \vdots \\ 2x_n x_2 \\ \vdots \\ x_n^2 \end{bmatrix} \in \mathbb{R}^{n(n+1)/2}, \tag{5.A.6}$$

and $*$ denotes the symmetric element.

**Remark 5.A.1:** In practice, $\tilde{\mathbf{x}}_t$ can be obtained by constructing

$$\tilde{\mathbf{X}}_t = \begin{bmatrix} x_1^2 & * & \cdots & \cdots & * \\ 2x_2x_1 & x_2^2 & \ddots & & \vdots \\ \vdots & \ddots & \ddots & \ddots & \vdots \\ \vdots & & \ddots & \ddots & * \\ 2x_nx_1 & \cdots & \cdots & 2x_nx_{n-1} & x_n^2 \end{bmatrix} \qquad (5.A.7)$$

first and taking the lower triangular part to form the vector.      ∎

At this stage, (5.A.4) can be rewritten as

$$\hat{x}_t = \tilde{\mathbf{h}}^{\mathsf{T}}\tilde{\mathbf{x}}_t + \mathbf{p}^{\mathsf{T}}\mathbf{x}_t + r$$

$$= [\tilde{\mathbf{h}}^{\mathsf{T}} \ \mathbf{p}^{\mathsf{T}} \ r] \begin{bmatrix} \tilde{x}_t \\ x_t \\ 1 \end{bmatrix} \qquad (5.A.8)$$

$$= \hat{\mathbf{h}}^{\mathsf{T}}\hat{\mathbf{x}}_t.$$

Consequently, (5.A.3) becomes

$$\min f(\hat{\mathbf{h}}) = \sum_{t=\zeta}^{T} (\hat{\mathbf{h}}^{\mathsf{T}}\hat{\mathbf{x}}_t - x_t)^2. \qquad (5.A.9)$$

Since this is an unconstrained optimization problem, we can simply take the gradient and make it to be 0 to obtain the global closed-form solution. Consequently, we have

$$f(\hat{\mathbf{h}}) = \sum_{t=\zeta}^{T} (\hat{\mathbf{h}}^{\mathsf{T}}\hat{\mathbf{x}}_t\hat{\mathbf{x}}_t^{\mathsf{T}}\hat{\mathbf{h}} + x_t^2 - 2\hat{\mathbf{h}}^{\mathsf{T}}\hat{\mathbf{x}}_tx_t)$$

$$= \hat{\mathbf{h}}^{\mathsf{T}}\left(\sum_{t=\zeta}^{T} \hat{\mathbf{x}}_t\hat{\mathbf{x}}_t^{\mathsf{T}}\right)\hat{\mathbf{h}} - 2\hat{\mathbf{h}}^{\mathsf{T}}\left(\sum_{t=\zeta}^{T} x_t\hat{\mathbf{x}}_t\right) + \sum_{t=\zeta}^{T} x_t^2. \qquad (5.A.10)$$

Then, the gradient can be calculated as

$$\nabla f(\hat{\mathbf{h}}) = 2\left(\sum_{t=\zeta}^{T} \hat{\mathbf{x}}_t\hat{\mathbf{x}}_t^{\mathsf{T}}\right)\hat{\mathbf{h}} - 2\sum_{t=\zeta}^{T} x_t\hat{\mathbf{x}}_t. \qquad (5.A.11)$$

Thus, $\nabla f(\hat{\mathbf{h}}) = 0$ implies

$$\hat{\mathbf{h}}^* = \left(\sum_{t=\zeta}^{T} \hat{\mathbf{x}}_t\hat{\mathbf{x}}_t^{\mathsf{T}}\right)^{-1} \sum_{t=\zeta}^{T} x_t\hat{\mathbf{x}}_t, \qquad (5.A.12)$$

which is a global closed-form solution of (5.A.4). As long as $\hat{\mathbf{h}}^*$ is obtained, $\mathbf{H}$, $\mathbf{p}$, and $r$ can be recovered from it.

**Remark 5.A.2:** In practice, there exists an overwhelming probability that the inverse of $\left( \sum_{t=\zeta}^{T} \hat{\mathbf{x}}_t \hat{\mathbf{x}}_t^{\mathsf{T}} \right)$ exists. This is because, on the one hand, the matrix is a symmetric matrix; on the other hand, since thousands of historical data will be used, the possibility of this matrix being singular is small. If occasionally the matrix is singular, we can simply add or remove one set of historical data to make the inverse exist. ∎

## 5.A.2 Quadratic ARMAX Model

Compared with the load, the temperature information is much easier to obtain. As a result, we can introduce the current and historical dry-bulb temperature as the exogenous input in the original forecasting model (5.A.1). The quadratic ARMAX model can be represented as

$$\hat{x}(t) = \mathbf{x}^{\mathsf{T}}(t)\mathbf{H}\mathbf{x}(t) + \mathbf{x}^{\mathsf{T}}(t)\mathbf{p} + \mathbf{z}^{\mathsf{T}}(t)\mathbf{G}\mathbf{z}(t) + \mathbf{z}^{\mathsf{T}}(t)\mathbf{q} + r, \tag{5.A.13}$$

where

$$\mathbf{z}(t) = \begin{bmatrix} z(t) \\ z(t-1) \\ \vdots \\ z(t-m+1) \end{bmatrix} \in \mathbb{R}^m \tag{5.A.14}$$

is the current and historical dry-bulb temperature and $\mathbf{G} \in S^m$, $\mathbf{q} \in \mathbb{R}^m$. Let $\hat{x}_t$, $\mathbf{x}_t$, $x_i$, $\mathbf{z}_t$, and $z_i$ denote $\hat{x}(t)$, $\mathbf{x}(t)$, $x(t-i)$, $\mathbf{z}(t)$, and $z(t-i+1)$, respectively. We have (5.A.13) rewritten as

$$\hat{x}_t = \mathbf{x}_t^{\mathsf{T}}\mathbf{H}\mathbf{x}_t + \mathbf{p}^{\mathsf{T}}\mathbf{x}_t + \mathbf{z}_t^{\mathsf{T}}\mathbf{G}\mathbf{z}_t + \mathbf{q}^{\mathsf{T}}\mathbf{z}_t + r. \tag{5.A.15}$$

In order to identify the model parameters $\mathbf{H}$, $\mathbf{p}$, $\mathbf{G}$, and $\mathbf{q}$, we still need to solve the optimization problem in the form of (5.A.3). By following similar lines as in Section 5.A.1, we can construct

$$\tilde{\mathbf{g}} = \begin{bmatrix} g_{11} \\ g_{21} \\ \vdots \\ g_{m1} \\ g_{22} \\ \vdots \\ g_{m2} \\ \vdots \\ g_{mm} \end{bmatrix} \in \mathbb{R}^{m(m+1)/2}, \tilde{\mathbf{z}}_t = \begin{bmatrix} z_1^2 \\ 2z_2 z_1 \\ 2z_3 z_1 \\ \vdots \\ 2z_m z_1 \\ z_2^2 \\ 2z_3 z_2 \\ \vdots \\ 2z_m z_2 \\ \vdots \\ z_m^2 \end{bmatrix} \in \mathbb{R}^{m(m+1)/2}, \tag{5.A.16}$$

and have the forecasting model represented as

$$\hat{x}_t = \tilde{\mathbf{h}}^\mathsf{T} \tilde{\mathbf{x}}_t + \mathbf{p}^\mathsf{T} \mathbf{x}_t + \tilde{\mathbf{g}}^\mathsf{T} \tilde{\mathbf{z}}_t + \mathbf{q}^\mathsf{T} \mathbf{z}_t + r$$

$$= [\tilde{\mathbf{h}}^\mathsf{T} \ \mathbf{p}^\mathsf{T} \ \tilde{\mathbf{g}}^\mathsf{T} \ \mathbf{q}^\mathsf{T} \ r] \begin{bmatrix} \tilde{\mathbf{x}}_t \\ \mathbf{x}_t \\ \tilde{\mathbf{z}}_t \\ \mathbf{z}_t \\ 1 \end{bmatrix} \tag{5.A.17}$$

$$= \hat{\mathbf{h}}^\mathsf{T} \hat{\mathbf{x}}_t.$$

Consequently, we can have the global closed-form solution in the same form as the solution for (5.A.3), which can be represented as

$$\hat{\mathbf{h}}^* = \left( \sum_{t=\zeta}^{T} \hat{\mathbf{x}}_t \hat{\mathbf{x}}_t^\mathsf{T} \right)^{-1} \sum_{t=\zeta}^{T} x_t \hat{\mathbf{x}}_t, \tag{5.A.18}$$

with $\zeta \geq \max\{m, n\}$.

**Remark 5.A.3:** The computational load of (5.A.12) mainly comes from the summation term, because in practice, $T - \zeta \geq 3 \times 10^4$. Compared with taking the matrix inverse, the summation takes a much longer time. However, this is not a big issue as the optimization is carried out off-line. If an on-line optimization is needed, the Sherman–Morrison formula can help to reduce the computational complexity. ∎

**Remark 5.A.4:** Normally, too many temperature lags do not help to improve the forecasting accuracy. Thus, in practice, only the current or 1-lag temperature is introduced in the model. ∎

## References

[1] F. Ding, Y. Shi, and T. Chen, "Performance analysis of estimation algorithms of nonstationary ARMA processes," *IEEE Transactions on Signal Processing*, vol. 54, no. 3, pp. 1041–1053, 2006.

[2] J. Y. Fan and J. D. McDonald, "A real-time implementation of short-term load forecasting for distribution power systems," *IEEE Transactions on Power Systems*, vol. 9, no. 2, pp. 988–994, 1994.

[3] S.-J. Huang and K.-R. Shih, "Short-term load forecasting via ARMA model identification including non-Gaussian process considerations," *IEEE Transactions on Power Systems*, vol. 18, no. 2, pp. 673–679, 2003.

[4] A. D. Papalexopoulos and T. C. Hesterberg, "A regression-based approach to short-term system load forecasting," *IEEE Transactions on Power Systems*, vol. 5, no. 4, pp. 1535–1550, 1990.

[5] S. Vemuri, W. L. Huang, and D. J. Nelson, "On-line algorithms for forecasting hourly loads of an electric utility," *IEEE Transactions on Power Apparatus and Systems*, vol. PAS-100, no. 8, pp. 3775–3784, 1981.

[6] M. T. Hagan and S. M. Behr, "The time series approach to short term load forecasting," *IEEE Transactions on Power Systems*, vol. 2, no. 3, pp. 785–791, 1987.

[7] I. Moghram and S. Rahman, "Analysis and evaluation of five short-term load forecasting techniques," *IEEE Transactions on Power Systems*, vol. 4, no. 4, pp. 1484–1491, 1989.

[8] D. C. Park, M. A. El-Sharkawi, R. J. Marks II, L. E. Atlas, and M. J. Damborg, "Electric load forecasting using an artificial neural network," *IEEE Transactions on Power Systems*, vol. 6, no. 2, pp. 442–449, 1991.

[9] J. W. Taylor and R. Buizza, "Neural network load forecasting with weather ensemble predictions," *IEEE Transactions on Power Systems*, vol. 17, no. 3, pp. 626–632, 2002.

[10] Y. Shi and H. Fang, "Kalman filter-based identification for systems with randomly missing measurements in a network environment," *International Journal of Control*, vol. 83, no. 3, pp. 538–551, 2010.

[11] H. M. Al-Hamadi and S. A. Soliman, "Short-term electric load forecasting based on Kalman filtering algorithm with moving window weather and load model," *Electric Power Systems Research*, vol. 68, no. 1, pp. 47–59, 2004.

[12] H. Fang, J. Wu, and Y. Shi, "Genetic adaptive state estimation with missing input/output data," *Proceedings of the Institution of Mechanical Engineering, Part I: Journal of Systems and Control Engineering*, vol. 224, no. 5, pp. 611–617, 2010.

[13] F. Palacios-Quiñonero, J. Vicente-Rodrigo, M. A. Molina-Hernández, and H. R. Karimi, "Improved switching strategy for selective harmonic elimination in DC-AC signal generation via pulse-width modulation," *Abstract and Applied Analysis*, vol. 2013, pp. 1–12, 2013.

[14] B.-J. Chen, M.-W. Chang, and C.-J. Lin, "Load forecasting using support vector machines: a study on EUNITE competition 2001," *IEEE Transactions on Power Systems*, vol. 19, no. 4, p. 2004–, 2004.

[15] B. Yu, H. Fang, Y. Lin, and Y. Shi, "Identification of Hammerstein output-error systems with two-segment nonlinearities: algorithm and applications," *Control and Intelligent Systems*, vol. 38, no. 4, pp. 194–201, 2010.

[16] Y. D. Song, Q. Cao, X. Du, and H. R. Karimi, "Control strategy based on wavelet transform and neural network for hybrid power system," *Journal of Applied Mathematics*, vol. 2013, pp. 1–8, 2013.

[17] H. Duan, J. Jia, and R. Ding, "Two-stage recursive least squares parameter estimation algorithm for output error models," *Mathematical and Computer Modelling*, vol. 55, pp. 1151–1159, 2012.

[18] Y. Liu, F. Ding, and Y. Shi, "An efficient hierarchical identification method for general dual-rate sampled-data systems," *Automatica*, vol. 50, no. 3, pp. 962–970, 2014.

[19] F. Ding, Y. Shi, and T. Chen, "Auxiliary model-based least-squares identification methods for Hammerstein output-error systems," *Systems & Control Letters*, vol. 56, no. 5, pp. 373–380, 2007.

[20] M. Liu, Y. Shi, and F. Fang, "Optimal power flow and PGU capacity of CCHP systems using a matrix modelling approach," *Applied Energy*, vol. 102, pp. 794–802, 2013.

[21] US Department of Energy. EnergyPlus. [Online]. Available: https://www.energyplus.net. Accessed February 7, 2017.

[22] IHS EViews. EViews. [Online]. Available: http://www.eviews.com. Accessed November 10, 2016.

# 6

# Complementary Configuration and Operation of a CCHP-ORC System

## 6.1 Introduction and Related Work

No matter what the current operating mode is, for an existing CCHP system, the electricity to thermal energy output ratio of the PGU is fixed. When the electricity to thermal energy requirement ratio of users is equal to the output ratio of the PGU, the exhausted waste heat by the PGU can just match the needs of cooling and heating, and the maximum efficiency of the CCHP system will be achieved. Unfortunately, this ideal situation rarely occurs because of the random variation of energy requirements. Stemming from different operating modes, the thermal energy or electricity supply of the PGU may be less or more than the requirements.

For the supply shortage situations, adding extra energy (gas or electricity) is a natural and direct treatment. However, for the exceeding demand situations, the available solutions are influenced by many factors. Specifically, to dispose surplus electricity, there are two common approaches: selling it back to the power grid or converting it to thermal energy. The former closely depends on the power market rules and the latter requires the installation of other thermal engines. For the latter approach, the electric chiller, as a highly efficient power to cooling conversion device, has attracted much attention. Besides the approaches discussed in the preceding chapters, Cardona and Piacentino [1] have presented a conceptual scheme and procedure based on thermoeconomics. They introduced this methodology to the design and optimization of a CCHP system in which a hybrid cooling subsystem is utilized to fulfill the requirement of users. Canova et al. [2] employed MILP to optimize an integrated trigenerative system with electric chillers.

Similarly, to make full use of excess thermal energy, there are also two possible choices: storing it in thermal storage units [3, 4] or converting it to electricity by some equipment. The former is called a direct but passive treatment, for it can only store or release thermal energy when conditions permit; the latter is called an indirect but active treatment, because it can utilize the excess thermal energy and dynamically adjust the electricity to thermal energy output ratio of the CCHP system. In terms of the conversion of heat to electricity, some recent research focused on the organic Rankine cycle (ORC) system. Mago *et al.* [5] analyzed energetic, economical, and environmental performances of a combined CHP-ORC system and compared them with those of a standalone CHP system in different climate zones. Al-Sulaiman *et al.* [6] compared the energetic performance of three trigeneration systems: SOFC-trigeneration, biomass-trigeneration, and solar-trigeneration. In all of them, the ORC was combined as the power generator. These above research studies show that ORC technology has advantages in the use of low-grade thermal energy, and its integration with polygeneration systems has good application prospects.

The motivation of this chapter is to improve the comprehensive performance of CCHP systems in a wide load range by optimizing the configuration and operation. To this end, a complementary CCHP-ORC system is configured, in which an ORC and an electric chiller are included. Furthermore, an optimal operation strategy for this system is presented. Using this strategy, the output of the electric chiller and the ORC can be adjusted dynamically to keep the system running at the matched status according to the energy requirements of users. The effectiveness of the proposed system and strategy are verified through case studies of a hypothetical building.

This chapter is organized in the following way. Section 6.2 details the CCHP-ORC system configuration. In Sections 6.3 and 6.4, optimal operation strategies are designed for normal load and overload cases, respectively. The evaluation criteria of the CCHP-ORC system are given in Section 6.5. Section 6.6 provides case study results to verify the effectiveness of the proposed system configuration and operation strategies. Section 6.7 summarizes this chapter.

## 6.2   System Configuration and Formulation

In order to provide sufficient energy and improve the flexibility of energy supply, an optimal CCHP-ORC structure with adjustable electricity to thermal energy output ratio is presented in Figure 6.1.

Compared with conventional configurations, an ORC is added as an important component. Actually, the salient features of this system are shown not only in its structure, but also in its unique operation strategy.

Assuming the minimum technical limits are relaxed and all the involved equipment can be adjusted over a wide operation range, the CCHP-ORC system can be formulated as in the following.

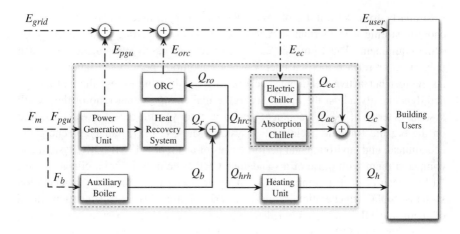

**Figure 6.1**    Structure diagram of a CCHP-ORC system

## 6.2.1    Energy Flows and Balances

If the system is running in a stable status and able to exactly match the energy requirements of users, referring to the energy flows shown in Figure 6.1, we can list energy
balance equations by the reverse derivation method.

### 6.2.1.1    Electricity Balance

The electric requirements of users, $E_{user}$, and the power input of electric chiller, $E_{ec}$,
can be supplied by the PGU, ORC, and the power grid as

$$E_{user} + E_{ec} = E_{pgu} + E_{orc} + E_{grid}, \qquad (6.1)$$

where $E_{pgu}$ and $E_{orc}$ are the electricity output of the PGU and ORC, respectively, and
$E_{grid}$ is the electricity provided by the power grid.

### 6.2.1.2    Thermal Energy Balances

The output of the absorption chiller, $Q_{ac}$, and/or the output of the electric chiller, $Q_{ec}$,
are applied to satisfy the cooling requirement of users, $Q_c$

$$Q_c = Q_{ac} + Q_{ec}. \qquad (6.2)$$

According to the changes of working conditions, recovered waste heat, $Q_r$,
is dynamically assigned to the heating unit, absorption chiller, and ORC. If the
recovered heat is not enough to cover these requirements, the auxiliary boiler will
start, leading to

$$Q_{hrh} + Q_{hrc} + Q_{ro} = Q_r + Q_b, \qquad (6.3)$$

where $Q_{hrh}$, $Q_{hrc}$, and $Q_{ro}$ are the heat inputs of the heating unit, absorption chiller, and ORC, respectively, and $Q_b$ is the heat output of the auxiliary boiler.

## 6.2.2 Equipment Efficiency and Energy Conversion

To fully describe the system's energy balance relationship, the efficiency coefficient and energy conversion equation of each piece of equipment are given below.

### 6.2.2.1 PGU, Heat Recovery System, and Auxiliary Boiler

In this chapter, the PGU efficiency is also represented by a second-order polynomial as shown in (3.7). With ORC and in normal cases, the function $f$ can be calculated as

$$f = \frac{\min\{E_{user} + E_{ec} - E_{orc}, \bar{E}_o^{pgu}\}}{\bar{E}_o^{pgu}}, \tag{6.4}$$

where $\bar{E}_o^{pgu}$ is the rated load of the PGU.

### 6.2.2.2 ORC

The ORC system involves the same components, including an evaporator, a turbine (expander), a condenser, and a pump, as in a conventional steam PGU. However, for such a system, the working fluid is an organic component characterized by a lower boiling point temperature than water and allowing reduced evaporating temperatures. ORC products have been on the market since the beginning of the 1980s. In recent years, manufacturers have provided ORC solutions over a wide range of power and temperature levels [7]. A schematic of a basic ORC system is shown in Figure 6.2.

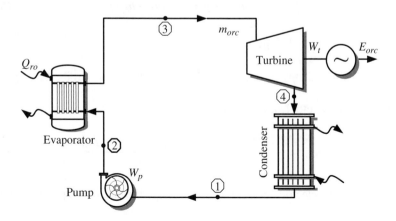

**Figure 6.2** Schematic of a basic ORC system

Four typical processes in the ORC are: pumping process ($1\rightarrow2$), heat addition process ($2\rightarrow3$), expansion process ($3\rightarrow4$), and heat rejection process ($4\rightarrow1$).

For the pumping process, the pump power, $W_p$, and isentropic efficiency, $\eta_p$, can be evaluated as

$$W_p = m_{orc}(h_2 - h_1) = \frac{m_{orc}(h_{2s} - h_1)}{\eta_p}, \tag{6.5}$$

$$\eta_p = \frac{h_{2s} - h_1}{h_2 - h_1}, \tag{6.6}$$

where $m_{orc}$ is the organic fluid mass flow rate, $h_1$ and $h_2$ are the enthalpies of organic fluid at the inlet and outlet of the pump, respectively, and $h_{2s}$ is the enthalpy at the outlet of the pump for the isentropic case.

Part of the recovered thermal energy from the heat recovery system of the CCHP system, $Q_{ro}$, is used for heating the organic working fluid in the evaporator. Therefore, the obtained heat, $Q_{ep}$, by the evaporator is

$$Q_{ep} = Q_{ro}\xi = m_{orc}(h_3 - h_2), \tag{6.7}$$

where $\xi$ is the evaporator effectiveness and $h_3$ is the enthalpy of organic fluid at the outlet of the evaporator.

The heat exchange of the condenser, $Q_{cd}$, can be expressed as

$$Q_{cd} = m_{orc}(h_1 - h_4), \tag{6.8}$$

where $h_4$ is the enthalpy of organic fluid at the outlet of the turbine.

The turbine energy output, $W_t$, and its isentropic efficiency, $\eta_t$, are given by

$$W_t = m_{orc}(h_3 - h_4) = m_{orc}(h_3 - h_{4s})\eta_t, \tag{6.9}$$

$$\eta_t = \frac{h_3 - h_4}{h_3 - h_{4s}}, \tag{6.10}$$

where $h_{4s}$ is the enthalpy of organic fluid at the outlet of the turbine for the isentropic case.

Then, the overall efficiency of the ORC is defined as

$$\eta_{orc} = \frac{W_t - W_p}{Q_{ep}} = \frac{(h_3 - h_{4s})\eta_t - (h_{2s} - h_1)/\eta_p}{(h_3 - h_2)} \tag{6.11}$$

Considering the electric generator efficiency, $\eta_{gen}$, the electric output of the ORC can be written as

$$E_{orc} = Q_{ro}\eta_{orc}\eta_{gen}. \tag{6.12}$$

## 6.2.3   Two Key Adjustable Parameters

In order to fully display the advantages of this CCHP-ORC system, two key adjustable parameters, $\alpha$ and $\beta$, are set. By adjusting these two parameters, the energy outputs of the electric chiller and ORC can be changed dynamically to ensure that the system works in the most economical operating condition while meeting the energy requirements of users.

The first adjustable parameter is the electric cooling to cool load ratio as detailed in Chapter 3. Herein, we use $\alpha$ to denote this ratio. Another adjustable parameter, $\beta$, reflects the ratio of the output of ORC to the total electricity requirement. It is expressed as

$$\beta = \frac{E_{orc}}{E_{user}}. \tag{6.13}$$

Since the PGU always provides part of the electricity, the range of reasonable values is $\beta \in [0, 1]$.

## 6.2.4   Electricity to Thermal Energy Output Ratio

When the PGU capacity is determined, the produced electricity and recoverable heat of the PGU satisfy

$$E_{pgu} = KQ_r, \tag{6.14}$$

where

$$K = \frac{\eta_{pgu}}{\eta_{hrs}(1 - \eta_{pgu})}. \tag{6.15}$$

For conventional CCHP systems without the electric chiller and the ORC, if there is no external energy inputs (such as the grid power and the supplementary thermal energy from the auxiliary boiler), the electricity to thermal energy output ratio of the CCHP system will be uniquely determined by the PGU. At that time, the CCHP system can achieve maximum efficiency only when the electricity to thermal energy requirement ratio of users is equal to $K$. However, this ideal condition is hard to reach in practice due to random changes of users' requirements.

Fortunately, by adjusting the loads of the electric chiller and the ORC dynamically, the electricity and thermal energy outputs of the presented CCHP-ORC system can be changed to accurately match the energy requirements of users over a wide load range. It means that the electricity to thermal energy output ratio of the new system is no longer only determined by the PGU, and the maximum efficiency of the whole system can be reached whether the electricity to thermal energy requirement ratio of users is equal to $K$ or not.

## 6.3   Optimal Operation Strategy for Normal Load Cases

If the rated capacity of the CCHP-ORC system is sufficient to satisfy the peak loads of users, that is, $E_{user} \leq \bar{E}_o^{pgu} + E_{orc\_r}$ and $Q_{eq} \leq \bar{Q}_r$, the working conditions can be segregated to three normal load cases based on the relationship between electricity and thermal energy requirements.

### 6.3.1   Normal Load Case 1: $E_{user} = KQ_{eq}$

This is an ideally matched condition for the CCHP-ORC system. Here, $Q_{eq}$ is the equivalent total thermal requirement of users at the output of heat recovery system, leading to

$$Q_{eq} = \frac{Q_h}{\eta_h} + \frac{Q_c}{COP_{ac}}. \tag{6.16}$$

In this case, the PGU can act in concert with the heating unit and absorption chiller to achieve the exact supply of users' requirements. The additional grid electricity and auxiliary heat are not needed, and the electric chiller and ORC are turned off. Accordingly, by setting $\alpha = 0$ and $\beta = 0$, the fuel consumption of the PGU is calculated as

$$F_{pgu} = \frac{E_{user}}{\eta_{pgu}}. \tag{6.17}$$

### 6.3.2   Normal Load Case 2: $E_{user} < KQ_{eq}$

This is one of the unbalanced conditions. In this case, the ORC should be turned off because of the heavy thermal load, that is $\beta = 0$.

#### 6.3.2.1   In the Adjustable Range

We first assume that the balance of energy requirements can be reached by only adjusting $\alpha$. Let $E_{orc} = 0$, $E_{grid} = 0$ and $Q_b = 0$, the electricity balance equation (6.1) becomes

$$E_{pgu} = E_{user} + \frac{\alpha Q_c}{COP_{ec}}. \tag{6.18}$$

By removing the heat input of ORC, the thermal energy balance equation (6.3) is transformed to

$$Q_r = \frac{Q_h}{\eta_h} + \frac{(1 - \alpha)Q_c}{COP_{ac}}. \tag{6.19}$$

Substituting (6.18) and (6.19) into (6.14), we have

$$E_{user} + \frac{\alpha Q_c}{COP_{ec}} = K \left[ \frac{Q_h}{\eta_h} + \frac{(1 - \alpha)Q_c}{COP_{ac}} \right], \tag{6.20}$$

where $Q_c$, $Q_h$, and $E_{user}$ can be collected from users in real time, and efficiency parameters are given by manufacturers. Thus $\alpha$ can be derived as

$$\alpha = \frac{K\left(\frac{Q_c}{COP_{ac}} + \frac{Q_h}{\eta_h}\right) - E_{user}}{\frac{Q_c}{COP_{ec}} + \frac{KQ_c}{COP_{ac}}}. \tag{6.21}$$

Furthermore, we can obtain the amount of fuel that should be supplied to the PGU as

$$F_{pgu} = \frac{E_{user}}{\eta_{pgu}} + \frac{\alpha Q_c}{COP_{ec}\eta_{pgu}} = \frac{K\left[E_{user}COP_{ec} + COP_{ac}\left(\frac{Q_c}{COP_{ac}} + \frac{Q_h}{\eta_h}\right)\right]}{(KCOP_{ec} + COP_{ac})\eta_{pgu}}. \tag{6.22}$$

### 6.3.2.2 Out of the Adjustable Range

Because all the coefficients in (6.21) are positive and the prerequisite is $E_{user} < KQ_{eq}$, the value of $\alpha$ must be larger than 0. According to the definition, $\alpha$ also should be less than or equal to 1. Therefore, the necessary condition of energy requirements can be derived from (6.21) based on $\alpha \leq 1$, having

$$Q_c \geq COP_{ec}\left(\frac{kQ_h}{\eta_h} - E_{user}\right). \tag{6.23}$$

As is commonly known, in winter, the cooling requirement of users is often low or even 0. If (6.23) cannot be satisfied, $\alpha$ will be greater than 1. At this time, $\alpha$ will be set as 1, revealing that all of the cooling requirement will be supplied by the electric chiller, and the absorption chiller will be stopped, and the energy input of each piece of equipment needs to be recalculated.

The most common strategy for this case is: let the PGU supply the electric requirements of users and electric chiller, and start the auxiliary boiler to make up the shortfall of thermal requirement. Such that

$$F_{pgu} = \frac{E_{user}}{\eta_{pgu}} + \frac{Q_c}{COP_{ec}\eta_{pgu}}, \tag{6.24}$$

$$F_b = \frac{Q_h}{\eta_h\eta_b} - \frac{F_{pgu}(1 - \eta_{pgu})\eta_{hrs}}{\eta_b}. \tag{6.25}$$

The total fuel consumption of the CCHP-ORC system is the sum of $F_{pgu}$ and $F_b$.

## 6.3.3　Normal Load Case 3: $E_{user} > KQ_{eq}$

This is another unbalanced condition. In this case, the electric chiller is unsuitable to be started to increase the already heavy electrical load, resulting in $\alpha = 0$. Then, the problem of matching energy requirements will be addressed by adjusting $\beta$ without grid electricity input.

Setting $E_{ec} = 0$, $E_{grid} = 0$ and substituting (6.13) into (6.1), we have

$$E_{pgu} = (1 - \beta)E_{user}. \tag{6.26}$$

Accordingly, the recovered heat meeting requirements can be derived as

$$Q_r = \frac{Q_c}{COP_{ac}} + \frac{Q_h}{\eta_h} + \frac{\beta E_{user}}{\eta_{orc}\eta_{gen}}. \tag{6.27}$$

Substituting (6.26) and (6.27) into (6.14) gives

$$(1 - \beta)E_{user} = K\left(\frac{Q_c}{COP_{ac}} + \frac{Q_h}{\eta_h} + \frac{\beta E_{user}}{\eta_{orc}\eta_{gen}}\right). \tag{6.28}$$

Then $\beta$ is calculated as

$$\beta = \frac{E_{user} - K\left(\frac{Q_c}{COP_{ac}} + \frac{Q_h}{\eta_h}\right)}{E_{user}\left(1 + \frac{K}{\eta_{orc}\eta_{gen}}\right)}. \tag{6.29}$$

Since all the coefficients in (6.29) are positive and the prerequisite is $E_{user} > KQ_{eq}$, the value of $\beta$ must be located in the range of 0 and 1, which is consistent with its definition.

Accordingly, the needed natural gas by the PGU is obtained as

$$F_{pgu} = \frac{(1 - \beta)E_{user}}{\eta_{pgu}} = \frac{K\left[E_{user} + \eta_{orc}\eta_{gen}\left(\frac{Q_c}{COP_{ac}} + \frac{Q_h}{\eta_h}\right)\right]}{(K + \eta_{orc}\eta_{gen})\eta_{pgu}}. \tag{6.30}$$

### 6.3.4 Decision-making Process

The decision-making process is summarized as a flowchart in Figure 6.3. Following this process, we can calculate the required fuel input of the PGU and issue it to the operation control system. Moreover, we can determine the ON/OFF status and the current output ratio, i.e., $\alpha$ and $\beta$, of the electric chiller and ORC in real time.

## 6.4 Operation Strategy for Overload Cases

Generally, the designed capacity of CCHP-ORC system is sufficient to meet the energy requirements of users. However, under extreme weather conditions or during major events, the peak loads may exceed the supply capacity. For these cases, different operation strategies should be adopted.

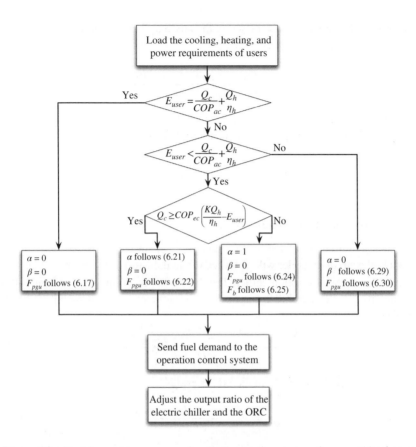

**Figure 6.3** Decision-making process of optimal operation strategy for normal load cases

### 6.4.1   Overload Case 1: $E_{user} > \bar{E}_o^{pgu}$ and $Q_{eq} > \bar{Q}_r$

This is an entire overload condition for the CCHP-ORC system. For this case, the PGU should be operated at rated load, $E_{pgu} = \bar{E}_o^{pgu}$. Accordingly, the fuel consumption of PGU will reach maximum $\bar{F}_{pgu}$

$$F_{pgu} = \bar{F}_{pgu} = \frac{\bar{E}_o^{pgu}}{\eta_{pgu}}. \tag{6.31}$$

Because the total thermal energy requirement $Q_{eq}$ is greater than the maximum recovered heat $\bar{Q}_r$, $\beta$ will be set to 0. According to the heat requirement $Q_h$, there are the following two sub-cases.

#### 6.4.1.1   Excessive Heat Requirement

If the heat requirement of the heating unit is greater than or equal to the recovered heat, that is, $\frac{Q_h}{\eta_h} \geq \bar{Q}_r$, the recovered heat will be fully delivered to the heating unit and the insufficient part will be offset by the auxiliary boiler. Consequently, the fuel consumption of auxiliary boiler is

$$F_b = \frac{Q_h}{\eta_h \eta_b} - \frac{\bar{F}_{pgu}(1 - \eta_{pgu})\eta_{hrs}}{\eta_b}. \tag{6.32}$$

At this moment, the cooling requirement will be entirely supplied by the electric chiller, that is, $\alpha = 1$. The purchased electricity from the power grid will be

$$E_{grid} = E_{user} + \frac{Q_c}{COP_{ec}} - \bar{E}_o^{pgu}. \tag{6.33}$$

### 6.4.1.2 Moderate Heat Requirement

If the heat requirement of the heating unit is less than the recovered heat, that is, $\frac{Q_h}{\eta_h} < \bar{Q}_r$, the auxiliary boiler will be turned off, and the rest of the recovered heat can be delivered to the absorption chiller. The difference in cooling requirement will be provided by the electric chiller due to its higher efficiency. Consequently, $\alpha$ can be calculated as

$$\alpha = \frac{Q_c - \left[\bar{F}_{pgu}(1 - \eta_{pgu})\eta_{hrs} - \frac{Q_h}{\eta_h}\right] COP_{ac}}{Q_c}. \tag{6.34}$$

The purchased electricity from the power grid is

$$E_{grid} = E_{user} + \frac{\alpha Q_c}{COP_{ec}} - \bar{E}_o^{pgu}. \tag{6.35}$$

## 6.4.2 Overload Case 2: $E_{user} > \bar{E}_o^{pgu}$ and $Q_{eq} \leq \bar{Q}_r$

Because of the excess of electric load, the PGU will be run up to full load and the electric chiller should be shut down. In this case, the recovered heat reaches its maximum $\bar{Q}_r$. Beside being supplied to the heating unit and the absorption chiller, the remaining heat is provided to the ORC as well. Therefore,

$$E_{orc} = \left[\bar{F}_{pgu}(1 - \eta_{pgu})\eta_{hrs} - \left(\frac{Q_h}{\eta_h} + \frac{Q_c}{COP_{ac}}\right)\right] \eta_{orc}\eta_{gen}. \tag{6.36}$$

Then, $\beta$ can be calculated as

$$\beta = \left[\bar{F}_{pgu}(1 - \eta_{pgu})\eta_{hrs} - \left(\frac{Q_h}{\eta_h} + \frac{Q_c}{COP_{ac}}\right)\right] \frac{\eta_{orc}\eta_{gen}}{E_{user}}. \tag{6.37}$$

The fuel consumption of the PGU is the same as (6.31) and the purchased electricity is

$$E_{grid} = (1 - \beta)E_{user} - \bar{E}_o^{pgu}. \tag{6.38}$$

## 6.4.3  Overload Case 3: $E_{user} \leq \bar{E}_o^{pgu}$ and $Q_{eq} > \bar{Q}_r$

In this case, the PGU can take all the electric load of users

$$E_{pgu} = E_{user} \tag{6.39}$$

and its fuel consumption is

$$F_{pgu} = \frac{E_{pgu}}{\eta_{pgu}} = \frac{E_{user}}{\eta_{pgu}}. \tag{6.40}$$

Since the recovered heat, $Q_r$, is not enough to meet the thermal energy requirement, the ORC has to be stopped. As discussed in Section 6.4.1, there are also two sub-cases.

### 6.4.3.1  Excessive Heat Requirement

If $\frac{Q_h}{\eta_h} \geq Q_r$, the additional fuel consumption of the auxiliary boiler will be

$$F_b = \frac{Q_h}{\eta_h \eta_b} - \frac{F_{pgu}(1 - \eta_{pgu})\eta_{hrs}}{\eta_b}. \tag{6.41}$$

The cooling requirement will be entirely supplied by the electric chiller, and the purchased electricity from the power grid will be

$$E_{grid} = E_{ec} = \frac{Q_c}{COP_{ec}}. \tag{6.42}$$

### 6.4.3.2  Moderate Heat Requirement

If $\frac{Q_h}{\eta_h} < Q_r$, the auxiliary boiler is no longer necessary, and the remaining recovered heat, $\frac{Q_h}{\eta_h} - Q_r$, will be sent to the absorption chiller. Apart from this, the insufficient part of the cooling requirement will be supplemented by the electric chiller

$$\alpha = \frac{Q_c - \left( F_{pgu}(1 - \eta_{pgu})\eta_{hrs} - \frac{Q_h}{\eta_h} \right) COP_{ac}}{Q_c}. \tag{6.43}$$

The energy input of the electric chiller still comes from the power grid

$$E_{grid} = E_{ec} = \frac{\alpha Q_c}{COP_{ec}}. \tag{6.44}$$

Based on the above analyses, the decision-making process for overload conditions can also be summarized by referring to Section 6.3.4.

## 6.5   EC Function of the CCHP-ORC System

The three criteria chosen to show the system performance are PEC, CDE, and COST. Definitions of PEC and CDE can be referred to in Chapter 3. COST in this chapter is defined as the daily cost as

$$COST = \left(F_{pgu} + F_b\right)\left(C_f + \mu_f C_{ca}\right) + E_{grid}C_e + \frac{R}{365}\sum_{j=1}^{m} N_j C_j, \qquad (6.45)$$

where $C_f$ and $C_e$ are the unit prices of natural gas and grid electricity, respectively, $C_{ca}$ is the unit carbon tax, $C_j$ is the initial capital cost of $j$th equipment, and $N_j$ is the installed capacity of $j$th equipment.

The last item of (6.45) is the daily capital cost for a total of $m$ equipment included in the CCHP-ORC system. Here, the capital recovery factor $R$ is defined as

$$R = \frac{i(1 + i)^n}{(1 + i)^n - 1}, \qquad (6.46)$$

where $i$ is the interest rate and $n$ is the equipment life cycle [8]. In this chapter, we assume that all equipment has the same service life.

It is evident that the lower PEC, CDE, and COST, the better performance of the CCHP-ORC system. And, in normal cases, there is always $E_{grid} = 0$.

To more clearly illustrate the advantages of the proposed CCHP-ORC system, we select conventional CCHP systems without electric chiller and ORC as the contrast system.

In order to avoid dealing with excess thermal or electric energy, the normal and popular operation strategy of the conventional CCHP system is to decide the output of the PGU according to the lower requirement between the $E_{user}$ and $KQ_{eq}$. The shortfall of another energy requirement will be supplemented by the auxiliary boiler or purchased from the grid [9]. So, in the criteria of the conventional CCHP system, electricity purchased from grid and fuel consumption of the PGU and auxiliary boiler will be all included. Despite the different structures and operation strategies, the above criteria calculation can also be used for the contrast system.

## 6.6   Case Study

### 6.6.1   Hypothetical Building Configuration

The configuration of this hypothetical building can be referred to in Section 2.5.1.

### 6.6.2   Test Results

A set of technical parameters of the CCHP-ORC system for the hypothetical hotel are listed in Table 6.1. For performance comparison, some parameters of a conventional CCHP system, conversion factors, and energy prices are also included.

**Table 6.1**  Technical parameters of the CCHP-ORC system and the CCHP system for the hypothetical hotel

| Symbol | Variable | Value |
|---|---|---|
| $a$ | First coefficient of $\eta_{pgu}$ (80 kW capacity) | −0.113 |
| $a$ | First coefficient of $\eta_{pgu}$ (100 kW capacity) | −0.153 |
| $b$ | Second coefficient of $\eta_{pgu}$ (80 kW capacity) | 0.282 |
| $b$ | Second coefficient of $\eta_{pgu}$ (100 kW capacity) | 0.316 |
| $c$ | Third coefficient of $\eta_{pgu}$ (80 kW capacity) | 0.086 |
| $c$ | Third coefficient of $\eta_{pgu}$ (100 kW capacity) | 0.088 |
| $\eta_{hrs}$ | Efficiency of the heat recovery system | 0.8 |
| $\eta_h$ | Efficiency of the heating unit | 0.8 |
| $\eta_b$ | Efficiency of the boiler | 0.8 |
| $COP_{ac}$ | COP of the absorption chiller | 0.7 |
| $COP_{ec}$ | COP of the electric chiller | 3.0 |
| — | Organic fluid | R113 |
| $m_{orc}$ | Mass flow rate (g/s) | 800 |
| — | Evaporator pressure (MPa) | 3 |
| — | Condenser temperature (°C) | 25 |
| $\eta_{orc}$ | Efficiency of the ORC | 0.199 |
| $\eta_{gen}$ | Efficiency of electric generator in ORC | 0.85 |
| $k_e$ | Site-to-primary electricity conversion factor | 3.336 |
| $k_f$ | Site-to-primary natural gas conversion factor | 1.047 |
| $\mu_e$ | $CO_2$ emission conversion factor of electricity (g/kWh) | 968 |
| $\mu_f$ | $CO_2$ emission conversion factor of natural gas (g/kWh) | 220 |
| $C_{ca}$ | Carbon tax rate (Yuan/kWh) | 0.00002 |
| $C_f$ | Natural gas rate (Yuan/kWh) | 0.19 |
| $C_e$ | Electricity rate at 6:00–21:00 (Yuan/kWh) | 0.93 |
| $C_e$ | Electricity rate at 22:00–5:00 (Yuan/kWh) | 0.55 |

### 6.6.2.1  Outputs of Electric Chiller and ORC

According to the energy requirements in representative days as in Figure 6.4 and the decision-making process shown in Figure 6.3, we can calculate the hourly outputs of the electric chiller and the ORC in representative days as in Figure 6.5. It is obvious that the electric chiller plays a significant role in summer, and the ORC has high utilization rate in the other three seasons.

### 6.6.2.2  Capacities of Equipment

Based on outputs of the electric chiller and the ORC in representative days, their capacities can be determined. After that, capacities of other equipment can be estimated for the CCHP-ORC system and the conventional CCHP system,

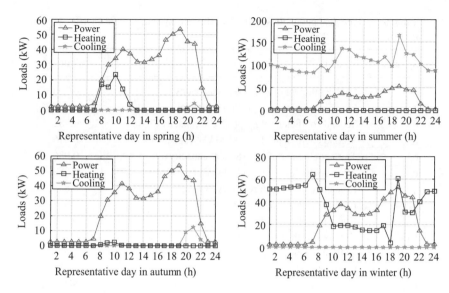

**Figure 6.4** Hourly cooling, heating and power loads of the hypothetical hotel in representative days of spring, summer, autumn, and winter

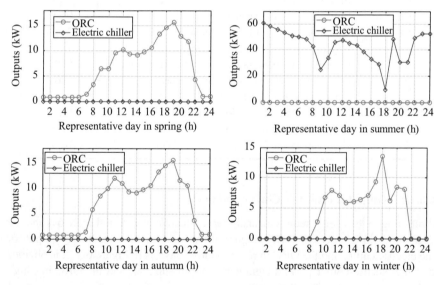

**Figure 6.5** Hourly outputs of the electric chiller and ORC in representative days of spring, summer, autumn, and winter

respectively. Referring to the current average price in the Chinese market, we can also give the corresponding unit prices for each piece of equipment. All of these data are listed in Table 6.2. In this example, the interest rate and service life of the equipment are set as $i = 6.21\%$ and $n = 20$ years, respectively.

**Table 6.2** Equipment capacities and unit prices of the CCHP-ORC system and the CCHP system

| Item | PGU | ORC | Boiler | Electric chiller | Absorption chiller | Heating unit |
|---|---|---|---|---|---|---|
| CCHP-ORC (kW) | 80 | 20 | 90 | 60 | 120 | 80 |
| CCHP (kW) | 100 | — | 170 | — | 180 | 80 |
| Unit prices (Yuan/kW) | 6800 | 12000 | 300 | 970 | 1200 | 200 |

**Table 6.3** Daily values of performance criteria for the CCHP-ORC system and the CCHP system in representative days

| Representative days | PEC (kWh) | | CDE (g) | | COST (Yuan) | |
|---|---|---|---|---|---|---|
| | CCHP-ORC | CCHP | CCHP-ORC | CCHP | CCHP-ORC | CCHP |
| Spring day | 1821.3 | 1974.8 | 382717.8 | 559706.2 | 504.9 | 756.2 |
| Summer day | 3986.7 | 5302.9 | 837711.6 | 1114270.8 | 990.3 | 1218.6 |
| Autumn day | 1801.1 | 1956.3 | 378449.4 | 561956.2 | 501.1 | 760.5 |
| Winter day | 2662.6 | 2801.2 | 559488.3 | 683277.8 | 734.0 | 863.9 |

### 6.6.2.3   Comparison and Discussion of System Performances

Upon acquiring the consumption of primary energy, the initial investment costs, and the equipment depreciation rate, it is reasonable to obtain the daily performance criteria (i.e., PEC, CDE, COST) for the CCHP-ORC and the conventional CCHP system as shown in Table 6.3. (With reference to the actual operation, the selected efficiency of the PGU in calculations is not less than 0.16.) In order to more visually compare the performance differences of the two systems in different seasons, the three criteria are integrated in radar charts as shown in Figure 6.6. It is clear that all the performances of the CCHP-ORC system are better (have smaller criteria values) than those of the conventional CCHP system. These superiorities benefitted from the electric chiller in summer and from the ORC in the other three seasons. Also, the total consumption in summer is significantly higher than that in the other seasons.

It should be noted that the cooperation of ORC and PGU can improve the energy efficiency and reduce the capacity requirements of the auxiliary boiler and PGU. Therefore, although the unit price is several times higher than for other equipment, the contribution of the ORC is greater than its cost.

## 6.7   Summary

A complementary CCHP-ORC system and its optimal operation strategy are presented in this chapter. The main feature of this system is that the electricity to

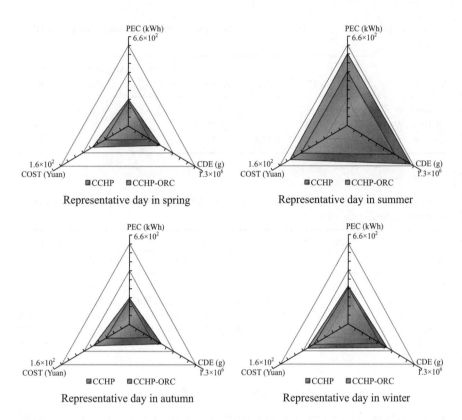

**Figure 6.6** Radar charts of three criteria for the CCHP-ORC system and the CCHP system in representative days of spring, summer, autumn, and winter

thermal energy output ratio can be adjusted dynamically according to the users' energy requirements. This adjustment is realized by changing the outputs of the electric chiller and the ORC based on an embedded decision-making process, which is derived from the optimal operation strategy. Case studies show the effects of this new configuration.

During the whole operation period, the energy requirements are supplied by the CCHP-ORC system in the matching status and no excess energy is produced. This is an ideal result, which avoids the policy constraint on electricity sold back and the loss of heat emission.

High investment cost of the ORC is the main problem of this system. But, as can be seen from the case study results, the ORC is fully utilized in spring, autumn, and winter. And, since the capacities of the PGU and the auxiliary boiler can be lowered due to the adding of the ORC, the initial installation cost of the entire system will be balanced to a certain extent. Moreover, the unit price of the ORC is continually declining on account of the technology maturing in recent years. Therefore, it is believed that the combination of the CCHP system and the ORC will be an attractive and valuable choice.

# References

[1] E. Cardona and A. Piacentino, "Optimal design of CCHP plants in the civil sector by thermoeconomics," *Applied Energy*, vol. 84, no. 7–8, pp. 729–748, 2007.

[2] A. Canova, C. Cavallero, F. Freschi, L. Giaccone, M. Repetto, and M. Tartaglia, "Optimal energy management-operational planning of an integrated trigenerative system," *IEEE Industry Applications Magazine*, vol. 15, no. 2, pp. 62–65, 2009.

[3] A. H. Azit and K. M. Nor, "Optimal sizing for a gas-fired grid-connected cogeneration system planning," *IEEE Transactions on Energy Conversion*, vol. 24, no. 4, pp. 950–958, 2009.

[4] S. M. Lai and C. W. Hui, "Feasibility and flexibility for a trigeneration system," *Energy*, vol. 34, no. 10, pp. 1693–1704, 2009.

[5] P. J. Mago, A. Hueffed, and L. M. Chamra, "Analysis and optimization of the use of CHP-ORC systems for small commercial buildings," *Energy and Buildings*, vol. 42, no. 9, pp. 1491–1498, 2010.

[6] F. A. Al-Sulaiman, F. Hamdullahpur, and I. Dincer, "Performance comparison of three trigeneration systems using organic Rankine cycles," *Energy*, vol. 36, no. 9, pp. 5741–5754, 2011.

[7] S. Quoilin and V. Lemort, "Technological and economical survey of organic Rankine cycle systems," in *Proceedings of European Conference on Economics and Management of Energy in Industry*, Vilamoura, Portugal, 2009, pp. 1–12.

[8] E. Cardona, A. Piacentino, and F. Cardona, "Energy saving in airports by trigeneration. Part I: Assessing economic and technical potential," *Applied Thermal Engineering*, vol. 26, no. 14–15, pp. 1427–1436, 2006.

[9] P. J. Mago and L. M. Chamra, "Analysis and optimization of CCHP systems based on energy, economical, and environmental considerations," *Energy and Buildings*, vol. 41, no. 10, pp. 1099–1106, 2009.

## References

[1] ...

# Index